BRUTAL REASONING

Brutal Reasoning

Animals, Rationality, and Humanity in Early Modern England

Erica Fudge

CORNELL UNIVERSITY PRESS
Ithaca and London

First published 2006 by Cornell University Press
First paperback printing 2019
Printed in the United States of America

Library of Congress Cataloging-in-Publication Data

Fudge, Erica.
    Brutal reasoning : animals, rationality, and humanity in early modern England / Erica Fudge.
        p. cm.
    Includes bibliographical references and index.
    ISBN-13: 978-0-8014-4454-8 (cloth : alk. paper)
    ISBN-10: 0-8014-4454-3 (cloth : alk. paper)
    ISBN: 978-1-5017-3087-0 (pbk. : alk. paper)
    1. Philosophical anthropology—History. 2. Human beings—Animal nature—History. 3. Animals (Philosophy)—History. 4. Reason—Philosophy—History. 5. Philosophy, English—16th century. 6. Philosophy, English—17th century. I. Title.
    BD450.F7944   2006
    128.0942—dc22   2006014459

For Tessa and Tim

God a mercy horse.

Contents

Acknowledgments

A book can only get written if there is time enough for writing, and my first acknowledgment must be to the various institutions that have allowed me the opportunity to begin, continue, and complete this project. I received a small grant from the British Academy in 1999 to get this book under way. I was granted teaching relief and a sabbatical from the School of Arts at Middlesex University, which enabled me to continue with my research. Finally a one-year research fellowship from the Leverhulme Trust gave me the much-needed time to finish. I am extremely grateful to all of these institutions for their support.

In addition to time, a writer needs colleagues and accomplices, and I have been extremely lucky to have many of these. I thank (as ever) Sue Wiseman and Lawrence Normand for reading drafts of bits of the book and for making helpful comments. My friends in the Animal Studies Group—Steve Baker, Jonathan Burt, Diana Donald, Garry Marvin, Robert McKay, Clare Palmer, and Chris Wilbert—have always been full of good ideas. Mary Baine Campbell, Brian Cummings, Peter Harrison, James Knowles, Chris Mounsey, Patricia Parker, Stella Sandford, Jonathan Sawday, and Alan Stewart have lent ears and brains in the course of the research and writing of this book.

Versions of parts of this book were given at conferences at Princeton University, University College Dublin, Aberdeen University, and the Oxford Renaissance Seminar; my thanks go to the respective organizers: Andrea Immel and Michael Whitmore, Jerome de Groot, Andrew Gordon, and Michelle O'Callaghan.

Staff at Cornell University Press have been a pleasure to work with, and

I thank in particular Nancy Ferguson, Teresa Jesionowski, Alison Kalett, and Bernhard Kendler. The manuscript greatly benefited from the sensitive copy-editing of David Schur. I am grateful to him for his exceptional work. Of course, any errors are mine alone.

I am grateful to the publishers for the right to reproduce the following:

Parts of chapters 2 and 3 appeared in "Learning to Laugh: Children and Being Human in Early Modern Thought," *Textual Practice* 17, no. 2 (2003): 277–94: www.tandf.co.uk.

Part of chapter 3 appeared in "Two Ethics: Killing Animals in the Past and the Present," in *Killing Animals,* ed. The Animal Studies Group (Urbana: University of Illinois Press, 2006), 99–119

E.F.

BRUTAL REASONING

Introduction

This book began its life as an attempt to examine what early modern English writers believed about the reasoning capacity of animals in the period before the appearance of Descartes's "beast-machine" hypothesis.[1] I wanted to see if any aspects of this hypothesis, that declared the automatism of animals and the absolute distinction of the human from the animal, could be traced in early modern English culture prior to the appearance of *Discourse on the Method* (1637). Did Descartes offer continuity or a radical shift in the ways in which English people thought about animals? Did the anthropocentrism of the early modern period find, as James Serpell has argued, its "inevitable and sinister conclusion" in the work of René Descartes,[2] or was the declaration of animal automatism a challenge to established ideas? And, finally, what impact did Descartes's ideas have on attitudes toward animals in England?

What emerged from the research I undertook was more complex, uneven, and of more central cultural significance than I had first expected. Discussions of what might broadly be termed the mental capacities of animals were present, and actually frequent, but they were usually to be found in dis-

---

[1] Throughout the book I use the term "animal" rather than "nonhuman animal." While much current work on animals prefers the latter terminology and regards it as simultaneously more accurate and respectful, the term "animal" follows early modern usage more closely. I also think that "animal" reflects popular modern usage, and that at times "nonhuman animal" can distance the reader from discussions in ways that are not always helpful: "animal" has an emotive quality that "nonhuman animal" lacks. My use of "man" and the masculine pronoun again reflects early modern usage. I discuss the gendering of reason in chapter 2.

[2] James Serpell, *In the Company of Animals: A Study of Human-Animal Relationships* (Oxford: Blackwell, 1986), 124.

cussions that were ostensibly concerned with human reason. Few, if any, treatises were written that specifically looked at the capacities of animals, and those that seemed to focus on this issue would slide into discussions of humans. In all of these works, whatever their apparent foci, the animal emerged as humanity's other; as the organism against which human status was asserted. However, animals were used not only in order to establish and reinforce human status in discussions of reason, or even *required* in such discussions. As well as this, it became clear that, as the human possession of reason was cited as the primary source of the difference between humans and animals in the early modern period, the centrality of animals in that discussion took on a more insistent and foundational power than I had initially supposed.

Because of this interrelationship between humans and animals in early modern English debates about reason, *Brutal Reasoning* looks at the ways in which humans were constructed, at what being human meant, and at how humans could lose their humanity. It also considers what made an animal an animal, why animals were looked at in the early modern period, and how people understood, and misunderstood, what they saw when they did look. In tracing this history *Brutal Reasoning* attempts to reevaluate some of the ways in which being human was represented and discussed in the period. It also aims to challenge some oversimplistic notions about early modern attitudes toward animals and about the impact of those attitudes in modern culture.

Despite the inseparability of humans and animals manifested in early modern discussions, however, there is an apparently unambiguous starting point: writers from classical times onward presented reason as the distinguishing feature of humanity. Early modern thinkers, following their intellectual forebears, appear clear on this. In 1631, for example, Daniel Widdowes stated,

> All Creatures are reasonable, or unreasonable. They which want reason, are Beasts, who live on Land or in Water. Those which live on the earth, moove on the earth, or in the ayre.[3]

In a very different vein, but with the same meaning, in Thomas Nashe's *Haue with you to Saffron-walden* (1596), Gabriel Harvey's tutor writes to Harvey's father to tell him a tale about his son:

> *he would needs defend a Rat to be Animal rationale, that is, to haue as reasonable a soule as anie Academick, because she eate and gnawd his bookes, and except she carried a braine with her, she could neuer digest or be so capable of learning.*

---

[3] Daniel Widdowes, *Natvrall Philosophy: Or A Description of the World, and of the severall Creatures therein contained,* 2nd ed. (London: T. Cotes, 1631), 64.

The play with language—the double meaning of *digest*—allows Nashe to simultaneously represent and comically undermine the philosophical distinction that Widdowes was to represent so very clearly. And this undermining of the opposition between reason and unreason means that it is somehow logical that Harvey, we learn later, libels the "*Colledge dog . . . onely because he proudly bare vp his taile as hee past by him.*"[4] Having overturned the primary difference between human and animal, Harvey cannot assert any others: a dog in a world without difference is as likely to be libeled as a man. And a human who sees no difference between a human and a dog is himself clearly acting like an animal.

In this way reason can be seen to play a crucial role in conceptualizations of the human and the animal. Without it, as the story of Harvey shows, all order and distinction would break down because ultimately discussions of reason in early modern England are discussions of order. Simply put, the human possession of reason places humans above animals in the natural hierarchy. Reason reveals humans' immortality, and animals' irrationality reveals their mortality, their materiality. Reasonable humans are the gods on earth. Because of the link between reason and natural order, then, texts that might appear to have little to say about the nature of animals become significant to the historian of animals. In discussing humans, their souls, their status, these texts are outlining the framework by which humans lived with, and declared dominion over, animals.

But the question of order that I am tracing is not only an issue in the early modern period; a modern notion of order comes to the fore as well. When modern readers read early modern texts we read to make sense of them, to contextualize them, perhaps, but still to make them speak to us in our own time. Part of our sense making is to maintain a notion of natural order— this may not always be the same natural order as can be found in the early modern period, but it remains an *order*—and we often do this by ignoring, or by making figurative, the animals present in early modern texts. Even in "posthumanist" readings, where the natural, given status of the human is questioned in relation to its construction in ideology, language, and performance, animals rarely appear, even when those early modern texts referred to by historians and literary critics of the period themselves seem to have much to say about beings other than humans. When Robert Burton, for example, argues that self-knowledge will help the readers of *The Anatomy of Melancholy* to know "how a Man differs from a Dogge" this can be interpreted literally in a way that fully responds to contemporary philosophical ideas.[5] In fact, I would argue that taking the dog figuratively is actually mis-

---

[4] Thomas Nashe, *Haue with you to Saffron-walden. Or, Gabriell Harueys Hunt is vp* (London: John Danter, 1596), sigs. K4ᵛ and Lᵛ.

[5] Robert Burton, *The Anatomy of Melancholy,* 2nd ed. (Oxford: Henry Cripps, 1624), 12.

reading Burton's statement. When we recognize the literal meaning of animals in numerous early modern texts, what emerges is a vast body of literature that is in fact concerned not only with humans but also with animals.

As well as this, and as a consequence, a pattern of thought that has been obscured in modern scholarship emerges. By taking seriously the animals in discussions of reason in early modern England—that is, by asserting that the animals within these texts are to be interpreted as animals and not simply as symbols of something else—then, *Brutal Reasoning* is also concerned to redress an imbalance in historical analysis. As with much other recent work in the history of animals, part of the project of this book is to recover animals from the silence of modern scholarship. Animals, I think, were taken seriously by the early modern people who wrote about and with them in their discussions, and I think we fail if we do not do likewise. In addition, however, the willingness to take the animals in these texts as animals is not merely a reflection of what is currently of interest to scholars; it is not just a reading back into the past the concerns of the present. Taking animals seriously also highlights the fact that the marginalization of animals in modern humanities research itself serves an important philosophical and moral function. It obliterates a way of thinking that raises questions about the nature of the animal and the human; that offers us another inheritance, another way of conceptualizing both ourselves and the world around us.

What follows, then, are seven chapters, each of which takes up a particular aspect of this debate. In the first chapter, I briefly outline the differences between humans and animals as they were understood in the intellectual orthodoxy that I am terming the discourse of reason. I then consider in more detail some of the properties of humanity, and particularly the relationship of the human to time in early modern ideas. This first chapter sets up the debates that the rest of the book focuses on. Challenges to the discourse of reason, which are to be traced in the early modern writings themselves and not only in my interpretation of them, are the focus of the later chapters.

In chapter 2, for example, I look at the question of the origin of human status in the human's biological life, and at the role of education in achieving human status. The questions asked here are, when does a human become a human? Are all humans human? Chapter 3 looks at the way in which the notion of humanity outlined in the previous two chapters was perceived by early modern writers to be undone through human vice. This chapter ultimately unearths a new early modern ethic that surrounded and challenged the Aristotelian order of things, and which offered a new way of thinking about the nature of humans and of animals.

The fourth chapter turns explicitly to animals and to the various ways in which they were interpreted in the period. Moving from the influence of

Plutarch to Aristotelian ideas, the chapter ends with an assessment of the impact on human–animal relations of the rediscovery of the skeptical writings of Sextus Empiricus in the sixteenth century. The fifth chapter attempts to apply all of these ways of conceptualizing animals to one particular animal: Morocco the Intelligent Horse. Famous in England in the late-sixteenth and early-seventeenth centuries, Morocco gave rise to much debate about the nature of his performances, and what can be traced in these debates is both a scholarly and a popular interest in questions about the status of animal reason.

The sixth chapter looks at the theory of the beast-machine, traces René Descartes's relationship to older ideas, and considers his impact in the mid- and late-seventeenth century. Finally, in the conclusion, I look at the use of Descartes in modern culture, and especially the place of Cartesianism in the writing of history.

My aim in *Brutal Reasoning* is to challenge the silent effacement of the animal. It is to underline the presence of the animal in early modern culture and to understand its absence in past and current scholarly work. In doing this I follow the lead of the early modern texts I refer to. That is, I begin with the human and, from the attempt to outline the nature of that apparently superior being, begin to trace the meaning and status of animals in early modern thought. In following this trajectory, I necessarily focus on humans in the first three chapters. But in looking at humans and their capacity to reason, the human perception of animals is given a new and vital context.

What is clear is that early modern writers were fascinated by animals to an extent that is surprising in relation to the relative absence of animals in modern critical interpretation of that period. Where modern commentators dismiss animals after an initial statement of difference (animals do not have rational souls), early modern writers continue to invoke and discuss them. It is, I think, only by bringing human and animal together, by reading one alongside the other, that we can hope to find any real meaning in the problems faced by both early modern and, I would argue, modern writers. When is a human a human, and when is an animal an animal? Such questions are probably doomed never to be answered definitively, but we can, I hope, attempt to come to a clearer understanding of the problem.

But before discussing the problems that are inherent in the discourse of reason, the challenges that were offered to it in the early modern period, the "solution" presented by Descartes, and the assessments and criticisms of that solution, however, this book begins with an overview of the fundamental principles of the discourse. Inevitably, because the discourse is so rife with problems, this overview should be taken as a preferred reading. Rather than challenging the discourse, the first chapter simply offers an analysis of its ideal outline of the nature of the human. So widely held was this discourse in early modern England that the texts used in the following discussion come from a range of areas—religious, philosophical, literary, even comic—from classical as well as early modern sources; from England and from other European nations. All the materials used were available to the English reader (and I am aware that at this time such a readership would have represented a small minority of the population, an issue I consider in later chapters), but all the different materials operate within what I am terming the discourse of reason, all offer explanations of the nature of the human that are consistent in their assumptions. What is also consistent is the prevalence of animals in these works. To explain the human, it seems, is to explain the animal; or perhaps that should be reversed: to explain the animal is to explain the human.

CHAPTER 1

Being Human

Put simply, early modern anatomists knew that the human body and the animal body were almost identical in the structure and overall workings of many of their organs. Even René Descartes advised his reader "to have the heart of some large animal with lungs dissected before him (for such a heart is in all respects sufficiently like that of a man)."[1] This, along with the moral strictures against using humans, was the reason why vivisectors used animals as their models on the slab. How else but through his vivisection of animals could William Harvey have discovered the circulation of the blood, when to see blood circulating required a live subject?

The body, then, was not a central source of difference, and even when the human physique was invoked to reiterate distinction this physical difference was always merely a sign of the other, more significant, mental division. For example, Plato's argument that, as Stephen Bateman put it in the late sixteenth century, "other beasts looke downeward to the earth. And God gaue to man an high mouth, and commaunded him to looke vp and beholde heauen" might seem to make the difference one of anatomy, but the implication of this anatomical dissimilarity is, to Bateman, clear: "& he gaue to men visages looking vpwarde towarde the starres. And also a mann shoulde seeke heauen, and not put his thought in the earth, and be obedient to the wombe as a beast."[2] Here the anatomical difference—man as biped—is

---

[1] René Descartes, *Discourse on the Method* (1637), in *The Philosophical Writings of Descartes*, translated by John Cottingham, Robert Stoothoff, and Dugald Murdoch (Cambridge: Cambridge University Press, 1985), 1:134.

[2] Stephen Bateman, *Batman vppon Bartholome, his Booke De Proprietatibus Rerum, Newly corrected, enlarged and amended* (London: Thomas East, 1582), sig. 12ᵛ.

used to make an intellectual, and moral, point. And that point—the human's capacity to contemplate the divine—relied on a difference that could not be found in the material worlds of anatomy and vivisection. It focused on the human possession of an inorganic essence, otherwise known as the rational soul.

This assertion of a physically invisible difference between humans and animals has a classical pedigree. The conventional early modern assumption about the soul comes, by way of the Christianizing of the Middle Ages, from Aristotle's *De anima,* and what is taken up by early modern thinkers is a distinction between different kinds of souls, and different kinds of beings. Aristotle's focus is on the living: "what has soul in it differs," he writes, "from what has not, in that the former displays life."[3] The living are plants, animals, and humans, and the nature of the ensoulment of these different groups explains where the properties that are particular to humanity find their source.

Invisible Differences

There are in the Aristotelian model three different kinds of soul—vegetative, sensitive, and rational—a trinity that is also discussed in terms of the binary of the organic (vegetative and sensitive) and the inorganic (the rational). The vegetative soul is shared by plants, animals, and humans and is the cause of nutrition, growth, and reproduction: all natural—unthought—actions. The sensitive soul is possessed by animals and humans alone (plants have only the vegetative soul) and is the source of perception and movement. The rational soul houses the faculties that make up reason—including will, intellect, and intellective memory—and is only found in humans. It is these faculties of the rational soul that are used to define the distinctive and superior nature of the human. To say that a human is a reasonable being, then, is to say that a human has a rational soul, is in possession of an immaterial essence that allows for the possibility of the exercise of reason. Animals cannot reason, so this argument goes; not because they are stupid or morally bankrupt but because they lack the essential faculty required for the exercise of reason. But humans and animals do share some aspects of behavior— those which are dictated by the sensitive soul.

The sensitive soul makes use of the "external senses"—sight, hearing, smell, taste, and touch—and the "internal senses"—common sense, fantasy, imagination, cogitation, and memory. These latter faculties are housed in

[3] Aristotle, *De anima* 413a20–25, translated by J. A. Smith, in *The Works of Aristotle* (Chicago: William Benton, 1952), 1:643.

the brain, and each faculty is situated in a particular area or ventricle of that organ. At the front of the brain is the common sense (ready to receive and bring together sense data from the individual external senses); in the middle the imagination, which brings the data of the individual senses together to make up a more complete perception; and at the rear the memory. So linked were the perceptual and physical capacities in early modern thought that the Italian memory theorist Gulielmus Gratorolus could propose that "it shalbe a token that they haue a good Memorie, whose hinder part of the head is great and longe: and they a weake Memorie, whose hinder parte of the head is as it were playne and equall with the necke."[4] Similarly, gestures could be said to disclose what kind of (invisible) thought process was taking place by revealing the area of the brain that was being used. In answer to the question, "*Wherefore is it, that when wee would conceive any thing, we put our hands to the forehead, and when we would call a thing to memory, wee scratch behinde the head?*" the author of a 1637 book of *Curiosities* answers, "By the reason of the diversity of the seates; for the Intellect is seated in the fore-part, the Memory in the hinder part, and the Fantasie in the interstice betweene them: and therefore by those actions we doe as it were summon each by a peculiar motion to the use of its function."[5] On this basis it is easy to see how these capacities are functions of the sensitive, that is, organic soul. Using these faculties, the animal and the human are able to perceive the external world of objects in their presence (with the external senses) or in their absence (through the internal senses of imagination, fantasy, and memory).

In *The Anatomy of Melancholy* Robert Burton offers an orthodox definition of the different physical sites of the organic soul: the liver, the heart, and the brain. Burton writes that these "noble" organs divide the body into three areas: the belly, the chest, and the head, and these three areas in turn represent three processes crucial to humans and animals: nutrition, feeling, and perception. The brain is for Burton the "most noble Organe vnder Heauen," and is "priuy Counsellor, and Chancellour to the *Heart*."[6] The passage between the three organs is made by the "animal spirits" (and "animal" here is used from its basis in the Latin *anima* [soul] rather than in reference to nonhuman creatures). The lowest types of spirits are the natural spirits that emerge during "the conversion of food into blood." These are taken around the body in the blood. The second level is the vital spirits, which are more refined natural spirits that are transformed by heat and air in the heart and lungs, and are sent around the body in the arteries. Finally, these vital spirits are themselves refined in the brain and sent out through the body's

[4] Gulielmus Gratorolus, *The Castel of Memorie: wherein is conteyned the restoring, augmenting, and conseruing of the Memorye and Remembrance* (London: Rouland Hall, 1562), sig. Biii^r.

[5] R. B., *Curiosities: Or The Cabinet of Nature* (London: N. and I. Okes, 1637), 199.

[6] Robert Burton, *The Anatomy of Melancholy* (Oxford: Henry Cripps, 1624), 18 and 15.

sinews, or nerves, as animal spirits. These are, Lily B. Campbell has noted, the "means by which the messages of the senses are sent to the brain."[7] Burton describes the nerves as "Membranes without, and full of Marrow within, they proceed from the Braine, and carry the Animall spirits for sense and motion."[8]

As well as being the site of the motive forces, the organic soul also houses the emotions. These emotions—also known as appetites or passions—were figured as "perturbations of the mind" by the English Jesuit Thomas Wright in 1601, perturbations that had the effect of "alter[ing] the humours of our bodies, causing some passion or alteration in them." According to the tradition that came down from Thomas Aquinas, there are eleven passions of two distinct kinds: concupiscible and irascible. The concupiscible or "coveting" passions are love, desire, delight, hatred, abomination, and sadness. The irascible or "invading" are hope, despair, fear, audacity, and ire. Wright goes on:

> God and Nature gaue men and beasts these naturall instincts or inclinations, to prouide for themselues, all those thinges that are profitable, and to auoyde all those things which are damnifieable: and this inclination may bee called, *Concupiscibili*, coueting; yet, because that GOD did foresee, that oftentimes there shoulde occurre impediments to hinder them from the execution of such inclinations, therefore he gaue them an other inclination [the irascible], to helpe themselues to ouercome or auoid those impediments, and to inuade or impugne whatsoeuer resisteth.[9]

Writing in 1620, the Dominican theologian Nicolas Coeffeteau termed these inclinations the "*desiring power* [and the] *Angry power: The one of which without the other sufficeth not for the health of the Creatures.*"[10] These two powers of the appetite were figured as natural by Wright, Coeffeteau, and Burton.

What follows from the arousal of passion—whether concupiscible or irascible—is motion. For Burton, "in vaine were it otherwise to desire and to abhorre, if wee had not likewise power to prosecute or eschue, by mouing the Body from place to place." In animals movement takes place, he writes, with the use of the "*Imagination* alone," whereas the "efficient cause" of movement in man is "*Reason,* or his subordinate *Phantasie,* which apprehends this good or bad obiect."[11] It is with movement, then, that the organic—per-

---

[7] Lily B. Campbell, *Shakespeare's Tragic Heroes: Slaves of Passion* (1930; repr., London: Methuen, 1961), 55–56.

[8] Burton, *Anatomy of Melancholy,* 14.

[9] Th. W. [Thomas Wright], *The Passions of the Minde* (London: V. S., 1601), 13–14, 41, and 39.

[10] Nicolas Coeffeteau, *A Table of Humane Passions* (London: Nicholas Okes, 1621), n.p.

[11] Burton, *Anatomy of Melancholy,* 24.

ception, passion—becomes evidently inorganic in the human. Because, as Gail Kern Paster has argued, passions "belonged to a part of the natural order jointly occupied by humans and animals,"[12] it is only when the external world has entered the internal world of the mind and been judged, and when judgment has led to the decision to act or not to act, that we witness the true realm of the human. It is in the judgment of the passion and in the subsequent choice as to whether the desire should be acted on or not that a human being emerges, inasmuch as this judgment and decision-making process takes place not in the sensitive but in the rational soul.

The rational soul, unlike the vegetative or sensitive soul, has no physical organ. Whereas the workings of the sensitive soul could be traced between the heart and the brain—the heart for passions, the brain for the perception of objects—the rational soul is immaterial and immortal, and it is, of course, in this immaterial essence that the human can be found: the rational soul is where the passions can, and should, be controlled; where the lustful urges of the body are judged; and where will overrides desire in order to produce the self-controlled and truly human human. This capacity is often represented negatively. For Thomas Wright the effect of unchecked passions is clear:

> By this alteration which Passions woorke in the witte, and the will, wee may vnderstand the admirable metamorphosis and change of a man from himselfe when his affects are pacified, and when they are troubled. *Plutarch* said they changed them like *Circes* potions, from men into beasts.[13]

By implication, following one's passions is an abandonment of reason and is to live like an animal with only the use of the sensitive soul.

But the immaterial soul is more than merely a control mechanism for the human: in his 1624 *Historie of the World* Sir Walter Raleigh proposes that humans have free will "whereas beasts, and all other creatures reasonlesse, brought with them into the World . . . and that euen when they first fell from the bodies of their Dams, the nature, which they could not change."[14] Humans in this vision have the freedom to choose, which ideally means they have the capacity to judge good from evil. Animals, on the other hand, are tied to their bodies and their natural inclinations; they can never change or improve themselves. And what is figured as good by an animal is good only on a material level. In "Epistle XCII" Seneca wrote "pleasure is the good of a beast," and expanded on this by asking "Doth the tickling of the body cause

---

[12] Gail Kern Paster, *Humoring the Body: Emotions and the Shakespearean Stage* (Chicago: University of Chicago Press, 2004), 135.

[13] [Wright], *Passions of the Minde,* 100.

[14] Sir Walter Raleigh, *The Historie of the World. In Five Bookes* (London: Walter Burre, 1614), 27.

a happy life?"[15] By implication, such is the "happiness" of an animal who can know what is good to eat but not what is "a good," an abstract concept. Such abstractions are only available through the use of reason and are therefore only available to humans.

Humans, though, are not perfect; their imperfection—corruption—was caused by the Fall. John Woolton, the Bishop of Exeter, wrote in 1576 that before the Fall, "In the hart all affections and appetitions did obey [man's] minde & will, neither was there dissētion in any thing." The struggle between the passions and the will is postlapsarian, and for Woolton, as well as, or perhaps because of, this dissention arising between the mind and the heart (reason and the body), "in his minde [man] loste the perfit knowledge of his God: So that always after it was full of darkenes, ignorance, folishnesse, and rebellion againste God." But such is the generosity of God that he

> hath left vnto [man] . . . *a power to discerne betweene things honest and vnhonest,* and to vnderstand the grounds of liberall artes, of good lawes, & of honest accions. This knowledge of reason, as the Philosophers call it, was not altogether extinct in mans ruyne. For it was Gods good pleasure, that there shoulde yet be some difference betweene reasonable man and brute beastes.[16]

The hierarchy of nature, then—the superiority of humans—comes in the discourse of reason from God, and humans exist as mind and body, as rational and corporeal beings. Raleigh, for instance, wrote that

> God created three sorts of liuing natures, (to wit) Angelicall, Rationall, and Brutall; giuing to Angels an intellectuall, and to Beasts a sensuall nature, he vouchsafed vnto Man, both the intellectuall of Angels, the sensitiue of Beasts, and the proper rationall belonging vnto man.[17]

The poet Richard Brathwait presented it slightly differently in 1630: "it is in God to know all things; in *Man* to know some things; in *Beast* to know nothing."[18]

However fallen and corrupt postlapsarian humanity might be, there remains a difference, a separation from animals. But the placement of humans between the angelic and the animal creates problems. The human is a self

[15] Seneca, "Epistle XCII," in *The Workes of Lvcivs Annævs Seneca, Both Morall and Naturall,* translated by Thomas Lodge (London: William Stansby, 1614), 385.

[16] John Woolton, *A Newe Anatomie of whole man, aswell of his body, as of his Soule* (London: Thomas Purfoote, 1576), fols. 12^v, 15^v, and 21^v–22^r.

[17] Raleigh, *Historie of the World,* 25.

[18] Richard Brathwait, *The English Gentleman* (London: Iohn Haviland, 1630), 76.

divided against itself, a constant struggle of mind against body, reason against desire. The aim is for the mind to gain and maintain the upper hand, to subdue the body, but the key metaphor to describe the relation of will and passion, which comes from Socrates' description of the human soul in Plato's *Phaedrus*, reveals the difficulty of the maintenance of reason:

> Let us then liken the soul to the natural union of a team of winged horses and their charioteer. The gods have horses and charioteers that are themselves all good and come from good stock besides, while everyone else has a mixture. To begin with, our driver is in charge of a pair of horses; second, one of his horses is beautiful and good and from stock of the same sort, while the other is the opposite and has the opposite sort of blood-line. This means that chariot-driving in our case is inevitably a painfully difficult business.[19]

Thomas Wright repeated this metaphor, and Lodowick Bryskett, friend of the poets Sidney and Spenser, wrote in 1606 of sense and feeling: "if these two vnruly and wild powers, which are the spring and fountaine head of all disordinate affections, be once wel tamed and broken, they do no lesse obey her cōmaundements, then the wel taught horses obey the coach-man."[20] Thomas Rogers, Chaplain to Richard Bancroft, the Archbishop of Canterbury, had likewise proposed in 1576 that lust "is a wylde and vnruly colt, & needeth a skilfull ryder: else wyl it breake other mennes hedges, and spoile good & vertuous plants."[21] But, ideally, the human would control his passions; would use reason to overwhelm the urges of the body. As such the human displayed his difference from animals, those beings which possessed no reasoning capacity and therefore were merely bodily. So, reason as a general concept is control, choice, superiority, but it is also evidenced in more particular capacities. The properties of reason are capacities which are available only to humans, and it is these that offer further insight into humanity's power and animal's lack.

Proper to Humans

The list of the properties of humanity, Jacques Derrida has argued, is vast. He includes nakedness, "speech or reason, the *logos*, history, laughing,

---

[19] Plato, *Phaedrus* 246a–b, translated by Alexander Nehamas and Paul Woodruff (Indianapolis: Hackett, 1995), 31.

[20] [Wright], *Passions of the Minde*, 30; Lodowick Bryskett, *A Discourse of Civill Life* (London: Edward Blount, 1606), 114.

[21] [Thomas Rogers], *A philosophicall discourse, Entituled, The Anatomie of the minde* (London: Andrew Maunsell, 1576), sig. 12[r].

mourning, burial, the gift, and so on." Such a list, he concludes, "can attract a non-finite number of other concepts, beginning with the concept of a concept."[22] In early modern England this list might be extended to include blushing, deliberate movement, dreaming, fear of death, free will, immortality, knowledge, knowledge of God, knowledge of good and bad, and moral philosophy. (The list could go on.) Derrida is right to note the absurdity of such a potentially endless list, but what is important to note is the primacy of reason in all of these properties of humanity. As Anthony Nixon wrote in 1616, reason is "the soueraigne part of the Soule," it is "Prince or Magistrate ouer the other partes, and vertues of the Soule."[23] Reason is what is evidenced in all of the early modern properties of humanity.

Even speech is understood to exist *after* reason: in early modern writings language is regarded as a way of expressing reason (and speech is evidence of language and therefore of reason); possession of a rational soul *precedes* and *allows* for language and speech.[24] The implication is that there is a difference between true speech and mere parroting. Parroting is a repetition of words without a comprehension of their meaning (hence a parrot is capable of this). Ralph Lever, writing in 1573, argued that human expression through language is evidence of something else: "Man firste doth conceyue trim deuises in his heade, and then (as the Poetes doe feigne of Jupiter) is pained as a womā in trauaile, till he haue vttered and published them, to be sene, & commended of others." He reiterates this point later in the same text, which is titled, helpfully, *The Arte of Reason, rightly termed, Witcraft:* "Words are voyces framed with hart and toung, vttering the thoughtes of the mynde."[25]

This priority of reason over language is put slightly differently by Thomas Wright. He argues that speech evidences the inner being of the speaker: too much speech reveals a fool—"the leaues of loquacitie spring from the roote of small capacitie"—while too little speech can reveal ignorance, but is often regarded as the sign of a thinker. Slow speech reveals slowness of wit and understanding; quick talkers have unsettled judgement: "they excell in apprehension, but faile in discretion." Affectation in speech "proceedeth from

[22] Jacques Derrida, "The Animal That Therefore I Am (More to Follow)," *Critical Inquiry* 28 (2002): 373–74.

[23] A. N. [Anthony Nixon], *The Dignitie of Man, Both in the Perfections of his Soule and Bodie* (Oxford: Ioseph Barnes, 1616), 36 and 49.

[24] Keith Thomas, however, has argued that speech is the primary site of the difference between humans and animals in the early modern period, and that the "second distinguishing quality [is] reason." *Man and the Natural World: Changing Attitudes in England, 1500–1800* (London: Penguin, 1984), 32.

[25] Ralph Lever, *The Arte of Reason, rightly termed, Witcraft, teaching a perfect way to argue and dispute* (London: Henrie Bynneman, 1573), sigs. *v^r and A^r.

a most vaine and notorious pride" and scoffing speeches "proceedeth from some, of meere simplicitie and foolishnes" or from "pride and enuie."[26] This relation of language to reason is given a comic twist by John Taylor, the Water Poet. He tells the story of a man who buys a supposedly multilingual jackdaw. Five weeks later, the bird has not uttered a word "otherwise than his fathers speech, *Kaw, Kaw.*" But instead of demanding his money back (a substantial sum of two French crowns), the owner of the bird sees silence as evidence of something else, something even more impressive: "though my Daw doe not speake, yet I am in good hope that he thinks the more."[27]

The essayist Sir William Cornwallis took the human capacity for speech a stage further in 1610, writing, "our preciousnes is reason, reasons seruant is speech, which is the messenger of reason, and reasons meditation: these are the cement of societies, to beare these with solitarines is to contend with nature for wisedom; who hath abilities, & vseth them not, as some creatures strength, others hornes what recken we them but brutish, and reasonles?"[28] In language, reason becomes "the cement of societies," and society—living communally, within rules—is human. Thomas Rogers reiterates the orthodox prioritization of reason when he writes that God "hath not onely endued [man] with reason and speeches, wherby he may both conceiue aswell that whiche hath beene, is, or may be, and also vtter and expresse that which is conceiued vnto others." However, he continues and links reason (and its expression in language) directly with dominion: God has also made man "a beholder of the Heauens, and not only hath indued him with reason, and with that gift of beholding the firmament (which no other creatures doo) but also hath brought all other things aswel senceles as hauing life: as Foules, fishes and foure footed beasts, vnder his power either to kil or keep them."[29]

But while language expresses the inner workings of the mind, other properties of humanity have a slightly different relation to reason, and it is a relation that is worth outlining in order to begin to see the real work that reason does in the human. If we turn to laughter, we can trace the distinction between the bodily and the reasonable, between the animal and the human, that is so crucial to the discourse of reason. We begin, as is often the case in these debates, with Aristotle.

---

[26] [Wright], *Passions of the Minde*, 166, 171, 174, and 176.

[27] John Taylor, *Taylor's Wit and Mirth*, in *Shakespeare Jest-Books: Reprints of the Early and very Rare Jest-Books supposed to have been used by Shakespeare*, edited by W. Carew Hazlitt (London: Willis & Southeran, 1864), 3:12.

[28] Sir William Cornwallis, "Of Solitarinesse and company," in *Essayes* (London: I. Windet, 1610), sig. Dd1^v–Dd2^r.

[29] [Rogers], *A philosophicall discourse*, sig. 162^r.

The Laughing Human

In *De partibus animalium* Aristotle famously wrote that man is "the only animal that laughs."[30] This statement became a commonplace for two millennia, repeated by numerous writers.[31] The reason that humans alone can laugh has for Aristotle an anatomical foundation. He writes that the midriff is present in all "sanguineous animals," and divides "the region of the heart from the region of the stomach." As such, this "partition-wall and fence" separates the "nobler from the less noble parts": the heart from the liver. As well as having this vital physiological function, the midriff also serves as a site of a further and more generally important division. Aristotle argues,

> That heating of [the midriff] affects sensation rapidly and in a notable manner is shown by the phenomena of laughing. For when men are tickled they are quickly set a-laughing, because the motion quickly reaches this part, and heating it though but slightly nevertheless manifestly so disturbs the mental action as to occasion movements that are independent of the will.

The shared physiology of man and animal with which Aristotle begins this section, breaks down with the impact of heating the midriff. He writes, somewhat tautologically, "That man alone is affected by tickling is due firstly to the delicacy of his skin, and secondly to his being the only animal that laughs."[32] What also breaks down is the importance of human reason in the establishment of the human as a separate and particular being. Will—a capacity of the rational soul that allows the human to judge, to decide whether to act with or against the passions—is overwhelmed by the body.

In early modern writings this bodily power of laughter is frequently repeated. My focus here is initially on a French text—Laurent Joubert's *Traité du Ris* (*Treatise on Laughter*) of 1579—but it is clear that many of the preoccupations of this treatise, the fullest discussion of laughter in the period, are also found in English writings. Following classical convention, Joubert, a physician, argues that laughter "is common to all, and proper to man." He outlines the physical effect of laughing:

> Everybody sees clearly that in laughter the face is moving, the mouth widens, the eyes sparkle and tear, the cheeks redden, the breast heaves, the voice be-

---

[30] Aristotle, *De partibus animalium*, 673a5, translated by William Ogle, in *Works of Aristotle*, 2:202.

[31] See M. A. Screech and Ruth Calder, "Some Renaissance Attitudes to Laughter," in *Humanism in France at the end of the Middle Ages and in the Early Renaissance* (Manchester: Manchester University Press, 1970), 216–28.

[32] Aristotle, *De partibus animalium* 672a15–673a5, pp. 201–2.

comes interrupted; and when it goes on for a long time the veins in the throat become enlarged, the arms shake, and the legs dance about, the belly pulls in and feels considerable pain; we cough, perspire, piss, and besmirch ourselves by dint of laughing, and sometimes we even faint away because of it.

Having concentrated on the physical response, he goes on to look beyond the apparent, visible manifestation of laughing to its internal, but still organic, operation. The pericardium, "the sheath or cover of the heart," he writes, "pulls on the diaphragm to which it is thoroughly connected in men, quite otherwise than in animals. . . . And this is (in my opinion) the reason, or at least one of the principal ones, why only man is capable of laughter." Once again, it is the human anatomy—this time the fact that in humans the pericardium is connected to the diaphragm—that causes laughter.

But, of course, laughter that is due to mere anatomy is not rational; it does not engage reason, and this is something that Joubert and other early modern writers were aware of. Being tickled may provoke laughter, but this is termed by Joubert "bastard laughter," as it is a purely bodily response requiring no operation of the rational soul (remember Seneca's notion of animal pleasure). The other "untrue" version of laughter is "dog laughter" or the "*cynic spasm*" as Joubert terms it. This is false, and emerges either through willed contortion of the face ("angry and threatening dogs have this look"), or injury (a knife to the diaphragm): it is therefore potentially both reasonable (willed), and unreasonable (bodily).[33] In the *Schoole of good manners* (1595) William Fiston reiterates Joubert's categories of bodily laughter by apparently conflating them: "some laugh so unreasonably," he writes, "that therewith they set out their teeth like grinning dogs."[34] The lack of reason and the overwhelmingly bodily response reduce the human to the status of an animal.

On this basis, the fact that laughter came to be regarded as a property of humanity would seem to be based on merely anatomical distinctions, and therefore not to involve the possession or lack of reason. But there is an alternative form of laughter that is still based on physiological factors—the possession of the link between pericardium and diaphragm—but that goes beyond the merely material. The true laugh (as opposed to the bastard or dog laugh) calls on the workings of the rational soul, and it is this laugh that is truly the property of humanity.

Joubert writes, following convention, that all passions "proceed . . . from the sensitive appetite," and that what follows the natural—and inevitable—

[33] Laurent Joubert, *Treatise on Laughter* (1579) translated by Gregory David De Rocher (Tuscaloosa: University of Alabama Press, 1980), 17, 28, 46, 47, and 76.

[34] W. F. [William Fiston], *The Schoole of good manners* (London: I Danter for William Ihones, 1595), sig. B8ʳ.

movement from perception to passion (from eye, to brain, to heart) is a struggle between rationality and sensuality; between reason and the body. The danger in this process is, of course, that reason will be overwhelmed by the motions of the heart; that the rational soul will be taken over by the sensitive. Joubert's image to illustrate this point comes, conventionally again, from Plato: the relation of the will to the heart, he writes, "is like a child on the back of a fierce horse that carries it impetuously about, here and there, but not without the child's turning it back some, and, reining in, getting it back on the path." He goes on to emphasize that despite this battle, reason can rule the heart through something akin to "civil or political [power] where with authority one points out obligations." The will, therefore, is "free . . . to choose or refuse the right thing." The power of *voluntary* (as opposed to natural or sensitive) movement is always at the command of the will, but the power of the sensitive appetite "does not obey immediately, and often contradicts the will, employing long arguments and various thoughts, after which it sometimes happens that the diverted will yields to the emotions." The fact that laughter can resist the will, that humans can laugh—and sometimes feel they *have* to laugh—in spite of themselves, shows that laughter cannot be wholly "under the rational . . . faculty": that it must be, Joubert writes, of the "sensitive faculty" which humans share with animals.[35]

However, it is because of the conflict between reason and passion that is innate to laughter that it becomes an important aspect of being human. In fact, a true laugh becomes a microcosmic exhibition of human-ness, and hence a key property of the human. Quentin Skinner has outlined an orthodox understanding of the nature of laughter in Renaissance thought. The precedent is again classical, but this time Socratic rather than Aristotelian.[36] According to Quintilian, laughter "has its source in things that are either deformed or disgraceful in some way."[37] In this he is following Socrates, who proposed that laughter is a "combined" emotion: "when we laugh at the ridiculous qualities of our friends, we mix pleasure with pain."[38] Joubert goes even further than his classical forebears: in the experience of joy, he argues, the heart expands, while in the experience of sorrow it shrinks.[39] One can die of joy; witness Gloucester's death in *King Lear* when his heart "Burst smilingly."[40] One cannot, however, die of laughter because laughter is a dilation

[35] Joubert, *Treatise on Laughter*, 32, 33, 34, and 35.

[36] Quentin Skinner, "Why Laughing Mattered in the Renaissance: The Second Henry Tudor Memorial Lecture," *History of Political Thought* 22, no. 3 (2001): 418–47.

[37] Quintilian, *Institutio oratoria,* cited in Skinner, "Why Laughing Mattered," 428.

[38] Plato, *Philebus,* 50a–b, translated by Harold N. Fowler (London: William Heinemann, 1962), 339.

[39] Joubert, *Treatise on Laughter,* 44

[40] William Shakespeare, *King Lear,* in *The Complete Works,* edited by Stanley Wells and Garry

and contraction of the heart, is a movement between joy and sorrow. For this reason laughter is perceived to proceed from "a double emotion": "laughable matter," Joubert writes, "gives us pleasure and sadness."[41]

This innately contemptuous quality of laughter—its mixture of pleasure and pain, in which recognition is tempered by difference (I too could trip over and fall on my backside, but, thankfully, have not)—uproots any links to joy that might be assumed to be inherent in this human expression. This connection between scornfulness and laughter is repeated by numerous early modern thinkers in England, including Thomas Wilson, Robert Burton (who of course takes on the persona of Democritus, the laughing philosopher in his *Anatomy of Melancholy*), Sir Thomas Browne, and Thomas Hobbes.[42] Joubert himself writes, "inasmuch as laughter is caused by something ugly, it does not proceed from pure joy, but has some small part of sadness."[43]

This is where we shift from viewing laughter as a bodily (and therefore unreasonable) phenomenon to laughter as something that invokes and requires reason: Laurel L. Hendrix, in fact, has called early modern laughter "an act of reading."[44] And this is the reason why laughter is proper to humans. It is not because the pericardium is not linked to the diaphragm in any animal other than the human; it is because only the human has the judgment that is required to truly laugh. We move from the bodily (animal) to the rational (human), and in this movement trace a distinction between passion and reason. But it is not merely the object of scorn that alerts observers to the presence of the true (as opposed to dog or bastard) laugh: the mode of expression—the nature of the laugh—is also important. A *loud* laugh, for example, does not show great reason: quite the contrary. The Reformed divine Richard Greenham wrote in 1601 that "a foole when hee laugheth liftes vp his voyce, but the wise man is scarse heard."[45] A wise man not only knows when it is appropriate to laugh, he also exercises control over his body; his almost silent laugh reflects the power of his mind. Christ, like Plato, never laughed.[46]

For this reason true laughter is a vital property of humanity. In laughter

---

Taylor (Oxford: Oxford University Press, 1988), 5.3.191. References to Shakespeare are to act, scene, and line in this edition unless otherwise noted.

[41] Joubert, *Treatise on Laughter,* 44.

[42] All are cited in Skinner, "Why Laughing Mattered."

[43] Joubert, *Treatise on Laughter,* 39.

[44] Laurel L. Hendrix, "'Mother of laughter, and welspring of blisse': Spenser's Venus and the Poetics of Mirth," *ELR* 23 (1993): 119.

[45] Richard Greenham, "Of Joy and Sorrow," in *The Workes of the Reverend and Faithfull Servant of Jesus Christ M. Richard Greenham* (London: Felix Kyngston, 1601), 350.

[46] Plato's lack of laughter is according to Diogenes Laertius, cited in Marjorie O'Rourke Boyle, "Gracious Laughter: Marsilio Ficino's Anthropology," *Renaissance Quarterly* 52, no. 3 (1999): 717.

mind and body are brought into potential conflict, but in true laughter the mind takes control of something that is possibly and powerfully out of its control. True laughter occurs when reason is not overwhelmed by the body, but actually overwhelms it. This is evidence—an enactment—of powerful reason. The laugh may be of the passionate animal body, but the true laugh is certainly of the reasonable human mind.

So in laughter a judgment is made—that something is wrong and therefore laughable—and this judgment is linked to what Robert Burton calls core capacities of human reason. For Burton the rational soul not only encompasses the vegetative and sensitive souls; it also has two key parts: "The *Vnderstanding*, which is the *Rationall* power, *apprehending*: the *will*, which is the *Rationall* power *mouing*, to which two, all the other *Rationall* powers are subiect and reduced." The will we have already seen at work in the discussion of laughter: the "understanding" is, for Burton, "*a power of the Soule, by which we perceiue, know, remember, and iudge as well Singulars, as vniuersals: hauing certaine innate notices or beginnings or arts, a reflecting action, by which it iudgeth of his owne doings, and examines them.*" Possession of understanding in turn offers, Burton writes, "three differences . . . betwixt a Man and a beast. As first, the sense only comprehends *Singularities,* the Vnderstanding *vniuersalities.* Secondly, the sense hath no innate notions: Thirdly, Brutes can not reflect vpon themselues."

Under the heading of "innate notions" Burton includes "*Synteresis,* or the purer part of *Conscience,* [which] is an innate Habit, & doth signifie a *consideration of the knowledge of the Law of God and Nature, to know good or euill.*"[47] In brief, this is an inseparable part of the rational soul, an aspect of what other early modern writers term "natural law." The Dutch legal theorist Hugo Grotius, for example, wrote in 1625 that "Natural Law is the Dictate of right Reason shewing moral turpitude, or moral necessity to be in some act, by its convenience or disconvenience with the Rational Nature: and consequently that it is forbidden or commanded by the Author of Nature, God."[48] In this way, possession of innate powers of judgment means not only that all humans should be capable of such judgment but also that such judgment is God-given, and that to act against it is to act against—or without—God. Reason, here, has a clear Christian foundation.

Where Burton differs from many other early modern discussions of the innate powers of the mind is in confining such powers to humans. For other thinkers, natural law is implanted in all creation. Sir Walter Raleigh, for example, argues that the Law of Nature is "that disposition, instinct, and for-

---

[47] Burton, *Anatomy of Melancholy,* 26, 27.

[48] Hugo Grotius, *De Jure belli ac pacis* (1625), translated as *The Illustrious Hugo Grotius of the Law of Warre and Peace* (London: T. Warren, 1654), 2–3.

mall quality, which God in his eternall prouidence hath giuen and imprinted in the nature of euery creature, animate, and inanimate. And as it
is *diuinum lumen* in Men, inlightning our formall reason; so is it more than
sense in Beasts; and more than vegetation in plants."[49] But while Burton and
other thinkers differ on this issue, what they agree on is that there is, in hu
mans, an innate capacity—inseparable from the rational soul—to be able
to, in Burton's words, "*know good or euill*"; to make, in short, moral judgments. These judgments come from understanding, and such understanding requires more than the sensitive soul can offer; it requires reason.

Time and Prudence

While the faculties of the organic soul are shared by humans and animals,
then, it is the faculties of the rational soul that distinguish them. For Raleigh
and the French Jesuit Nicolas Caussin these faculties are understanding, will,
and memory; for Sir John Davies, writing in 1599, they are wit, will, understanding, and memory; for Burton wisdom, judgment, memory, and reason.[50] But while the terminology and emphasis of individual thinkers
differs, one thing remains clear: in early modern England humans were perceived to be capable of moving beyond the material world, and of making
discriminating judgments. The source of these capacities is, once again, the
rational soul.

In 1612 William Jewel proposed that "reason abstracteth from visible
things, things inuisible, from corporeall incorporeall, things secret and mys-
ticall from such as are plaine and triuiall, and lastly things generall from
things particular."[51] Likewise, Sir John Davies wrote of the human soul:

> From their grosse *matter* she abstracts the *formes*,
>   And drawes a kind of *Quintessence* from things,
>   Which to her proper nature she transformes,
>   To beare them light on her celestiall wings;
> This doth she, when from things *particular,*
>   She doth abstract the *vniuersall kinds,*
>   Which bodilesse, and immateriall are,
>   And can be lodg'd but onely in our minds;

[49] Raleigh, *Historie of the World,* 228.

[50] Ibid., 19; Nicolas Caussin, *The Holy Covrt, or The Christian Institution of Men of Qvality,*
translated by T. H. [Thomas Hawkins] (Paris: St. Omer, 1626), 304; John Davies, *Nosce Teipsum* (London: Richard Field, 1599), 47, 49–50; Burton, *Anatomy of Melancholy,* 18.

[51] William Jewel, *The Golden Cabinet of true Treasure: Containing the summe of Morall Philosophie* (London: John Crostley, 1612), 55.

> And thus from diuerse *accidents* and *acts,*
> Which do within her obseruation fall,
> She goddesses, and powres diuine abstracts,
> As *Nature, fortune,* and the *vertues* all.[52]

Where humans can find the divine in the material, animals see and know only what is before their eyes. For an animal, the material world remains material. In pseudo-Aristotle's *Problemes* (1597) it is stated that "mankinde onely, is partaker of supernaturall things, other beasts haue regard vnto that onely which they see."[53]

This difference between humans and animals manifests itself in early modern debates in one way in particular: the notion of abstraction is related to the capacity to live beyond the present. In emphasizing this, early modern writers are following established convention: Seneca argued "For time consisteth of three parts, of that which is past, of that which is present, of that which is to come. That which is onely present and shortest, and passeth soonest is giuen to beasts: as touching that which is past, they haue eyther none or little remembrance thereof neyther, but casually thinke they on things that are present."[54] Saint Augustine reiterated this, writing that "the interpretation of the past and future from the present" "belongs to man exclusively."[55]

In the early modern period this idea was repeated by numerous writers. Thomas Wright stated,

> Though men and beasts in many things differ, yet in one we may most plainely distinguish them, for beasts regard only or principally what concerneth the present time, but men forecast for future euents; they knowe the meanes and the end, and therefore comparing these two together, they prouide present meanes for a future intent.[56]

Miles Sandys, Member of Parliament for Cambridge University, wrote in 1634 that "Betweene man and beast this is a speciall difference, that a Beast, onely as farre as hee is moved by sense, applyeth himselfe to that alone, which is present, very little perceiving a thing past, or to come."[57] Ten years

---

[52] Davies, *Nosce Teipsum,* 24.

[53] *The Problemes of Aristotle, with other Philosophers, and Phisitions* (London: Arnold Hatfield, 1597), sig. I[v].

[54] Seneca, "Epistle CXXIIII," in *Workes of Lvcivs Annævs Seneca,* 492.

[55] St. Augustine, *The Greatness of the Soul,* translated by Joseph M. Colleran (Westminster, Md., and London: Newman Press and Longmans, 1949), 100–101.

[56] [Wright], *Passions of the Minde,* 312.

[57] Sir Miles Sandys, *Prudence, The first of the Foure Cardinall Virtues* (London: W. Sheares, 1634), 72.

earlier Edward, Lord Herbert of Cherbury, had likewise written that "our faculties are concerned not only with the present but also with the past and future."[58] And Grotius wrote that "Man above other Animals, hath . . . judgement to discern what things delight or hurt, not present onely, but future."[59] Raleigh placed another emphasis on this human capacity: "wee our selues," he wrote, "account such a man for prouident, as, remembring things past, and obseruing things present, can by iudgement, and comparing the one with the other, prouide for the future, and times succeeding."[60] That is, the capacity to look beyond the present is not regarded as a virtue in itself: it is, rather, merely a capacity, but a capacity that allows humans to act virtuously; to act, in fact, with prudence.

William Jewel's definition of prudence likewise confirms the link between this virtue and the human capacity to look back to the past and contemplate the future:

> The morall Philosophers haue attributed vnto Prudence, the hauing of three eyes; memory, vnderstanding & prouidence. With the first eye, she considereth times past. With the second, time present; and with the third, that which is to come. Likewise, the Prudent man, by the consideration of things past, and that which hath ensued them, will iudge so (in the like cases) of that which is to come: and then with long (but not seeming tedious) deliberation, will attend the time, consider the perills, vnderstand the occasions: and giuing sometimes place vnto time, and alwaies to necessity (still prouided it be not repugnant to his dutie) wil at length, courageously set his hand vnto the worke. For, the remembrance of things past, to the Prudent man, standeth in great steade concerning things present, and also makes him likewise to foresee the things that are to come.[61]

Because it engages abstraction, judgment, and the transcendence of the material, prudence is regarded as chief among the four cardinal virtues (the three others are temperance, fortitude, and justice). "Yea, prudence," wrote Caussin, "holdeth [those other virtues] (as it were) enfolded in the plaites of her robe, and vnfoldeth them, according to place, tyme, persons, occasions, which to know is to know all."[62] But in order to be prudent the human must also possess the *means* to prudence; that is, the capacity to shift between past, present, and future. And there are logically three linked facets

---

[58] Edward, Lord Herbert of Cherbury, *De Veritate* (1624), translated by Meyrick H. Carré (Bristol: University of Bristol Press, 1937), 113.

[59] Grotius, *Law of Warre and Peace*, sig. *4ᵛ.

[60] Raleigh, *Historie of the World*, 15.

[61] Jewel, *Golden Cabinet*, 33–34.

[62] Caussin, *Holy Covrt*, 349.

of this capacity; memory, self-knowledge, and foresight. In discussions of each of these aspects of prudence the status of the human is underlined. What is significant is that prudence brings these together. It is for this reason that prudence is viewed as, in Lodowick Bryskett's words, "the soule of the mind."[63] It is the most reasonable part of reason; it is, if you like, the essence of reason.

A Memory of Humanity

In the discourse of reason there were understood to be two different kinds of memory: the first, sensitive memory, is shared by humans and animals, and the second, intellective memory, is available only to humans. The difference between these two kinds of memory reiterates that animals exist only in the present, a distinction first made by Aristotle in *De memoria et reminiscentia:*

> Recollecting differs from remembering not only in respect of the time, but in that many other animals share in remembering, while of the known animals one may say that none other than man shares in recollecting. The explanation is that recollecting is, as it were, a sort of reasoning. For in recollecting, a man reasons that he formerly saw, or heard, or had some such experience, and recollecting is, as it were, a sort of search. And this kind of search is an attribute only of those animals which also have the deliberating part. For indeed deliberation is a sort of reasoning.[64]

Guliemus Gratorolus—using different terminology—writes an early modern version of this:

> Aristotle thoughte good, to assigne two actes of Memoratiō: to wytte, Memorie and Remembraunce: although Remembraunce perteyneth to those thinges whiche we haue forgotten, and is the offyce of the extymatyue or cogitatyue vertue, not principally of the Memoratiue . . . or you may name that faculty to be the minde and vnderstandinge as Themistius saythe: because there is no power or facultie perceiued to wander about, but ŷ vnderstanding. And this w'out ŷ presence of ŷ obiecte is onely in Man: for with the presence of the obiect it is also founde in brute Beastes, as Aristotle hath as-

[63] Bryskett, *Discovrse of Civill Life*, 187.
[64] Aristotle, *De memoria et reminiscentia* 453a4, in *Aristotle on Memory*, by Richard Sorabji (London: Duckworth, 1972), 59.

sented, and as it euidentlye appearethe in a Greyhounde or Spayniel: and it is called the phantasticall sense.[65]

The difference that Gratorolus draws shows that animals require the object before them to perform an act of remembrance—the sheep needs a wolf to appear in front of it to remember that it fears wolves. This remembrance uses the animal's imagination. As Seneca wrote, "A dumbe beast . . . remembreth those things that are past at such time as that which awakeneth the sense, awakeneth it selfe, as a horse remembreth himselfe of his way when he is set into the beginning of it: whilst he standeth in the Stable he hath no remembrance thereof, although he hath trode it ouer many times."[66] Humans, on the other hand, can forget—lose the memory—and recollect by an act of reasoning in the absence of a sensory prompt: they can recall the way home because their act of recollection is not mere remembrance, but is understanding. On one level, then, memory is a "storehouse," a vault for previous experience. It is where things from the past are kept for future usage, with their presence in the storehouse prompted into action by external sense data. On another level, however, memory is an outcome of understanding. Memory cannot be fully characterized in spatial terms because it sometimes reclaims that which is lost, that which takes up no space, and this reclamation, of course, takes place in the inorganic, rational soul.

But humans have not only the capacity to remember but also have the potential to forget. As the Bishop of London, Henry King, noted in 1625, there is not "in the whole masse of Creation . . . so forgetting a Creature as *Man*." The reason for this is simple: "How well were it for Man-kinde, if we might glorie in that infirmitie which beasts may doe, they cannot be sayd to haue lost what they neuer had, nor to forget what they neuer had organs to remember. We had a great deale lesse sin to answer for, could we say so too."[67] Animals, this argument proposes, are naturally ignorant (although King's statement that they "neuer had organs to remember" is not strictly correct: animals do have sensitive memory, and the "organ" that they lack is in fact inorganic). Humans, however, have free will and so can make poor choices as well as good ones. They can remember or forget, can laugh like a human or like a dog—the decision is in the hands of the individual.

The capacity for reminiscence in the human facilitates more than merely organized living, however; it also aids the spiritual life. John Donne, for ex-

---

[65] Gratorolus, *Castel of Memorie*, sigs. Fv^v–Fvi^r.

[66] Seneca, "Epistle CXXIIII," in *Workes of Lvcivs Annævs Seneca*, 492.

[67] Henry King, *A Sermon Preached at White-Hall in Lent. 1625*, in *Two Sermons Preached at White-Hall in Lent, March 3. 1625 and Februarie 20. 1626* (London: John Haviland, 1627), 2 and 5–6.

ample, stated in a 1619 sermon that "if the memory doe but fasten vpon any of those things which God hath done for vs, that's the nearest way to him." He continues, "Tho the memory be placed in the hindermost part of the braine, defer not thou thy remembring to the hindermost part of thy life, but doe it now & *nunc in die,* now whilst thou hast light."[68] This is an image that is repeated by Henry King six years later.[69] For both men salvation is linked to the capacity to shift between past (what has been done), present (what is being done), and future (the life to come). To use one's intellective memory (although figuratively represented by Donne and King as found in an organ) is, in fact, to worship God.

Another link between rationality and immortality is made by William Hill in 1605 when he writes that "there is nothing in [earthly] natures which may haue the force of memorie, vnderstanding, or imagination; which remembreth thinges past, foreseeth things to come, and apprehendeth thinges present. Which thinges are onely diuine; neither can it be found from whence it should proceed, but from God."[70] The possession of a rational soul links the human directly to notions of divinity. Humans are not only reasonable, they are also immortal, but human immortality is not just a prize, it is also a responsibility, and part of the responsibility of all individuals is not only to use memory it is also to know themselves. Again, self-knowledge is an aspect of reason that distinguishes humans from animals.

Self Knowledge

"Thou thy selfe art the subiect of my Discourse." So begins Burton's *Anatomy of Melancholy.* Burton's reason for writing the *Anatomy* is at once coy—"I write of Melancholy, by being busie to avoid Melancholy"—and conventional. Burton offers two reasons for his *"Digreßion"* into the description of the human anatomy: first, because such a *"Digreßion"* is necessary to the clarity of the rest of the book, and second,

> it may peraduenture giue occasion to some men, to examine more accurately, and search farther into this most excellent subiect, that haue time and leasure enough, and are sufficiently informed in all other worldly businesses; as to make a good bargaine, buy, and sell, to keepe & make choise of good

[68] John Donne, *Sermon of Valediction at his Going into Germany Preached at Lincoln's Inn April 18, 1619* (London: Nonesuch Press, 1932), 15–16.

[69] King, *Sermon Preached at White-Hall,* 30–31.

[70] William Hill, *The Infancie of the Soule: Or, The Soule of an Infant* (London: C. Knight, 1605), sigs. B^v–B2^r.

Hauke, Hound, Horse, &c. but for such matters as concerne the knowledge of themselues, they are wholy ignorant and carelesse, they knowe now what this Bodie and Soule are, how combined, of what parts and faculties they consist, or how a Man differs from a Dogge.[71]

The final point is, inevitably, the crucial one here. It might seem an overstatement on Burton's part to assert that without knowledge of anatomy readers cannot know themselves, and therefore cannot tell the difference between a man and a dog, but what underlies this statement is utterly logical. Without self-knowledge a human is living a life without use of the rational soul; is living, therefore, the life of an animal. And the incapacity to differentiate human from dog represents a failure to exercise the difference. Miles Sandys puts this point in a similar way: "Doth the horse know that he is a horse, or, that he is a beast, and thou a man? . . . or doth the Dogge (which of all beasts is mans chiefe attendant) know, whether thou art a man, or a beast? no certainely." He continues: "onely man knowes that hee is man, and every beast in his severall kind, according to that of *Socrates;* Wisedome is in man, and not in a beast, and all wisedome is concluded in him in this word, [*Nosce teipsum*]."[72]

Know Thyself. This was the dictum that came from the Oracle at Delphi, was a staple in classical thought, and was likewise central to early modern ideas. Sir John Davies regarded this as the central question of his analysis of the human entitled, indeed, *Nosce Teipsum.*

> For how may we to other things attaine,
>> When none of vs his owne soule vnderstands?
>> For which the Diuell mockes our curious braine,
>> When *know thy selfe* his oracle commands.
> For why should we the busie Soule beleeue,
>> When boldly she concludes of that, and this,
>> When of herselfe she can no iudgement geue,
>> Nor how, nor whence, nor where, nor what she is?
> All things without, which round about we see,
>> We seeke to know, and haue therewith to do:
>> But that whereby we *reason, liue, and be,*
>> Within ourselues, we strangers are thereto.

This recognition of the need for knowledge of the self leads Davies to a logical conclusion: "My selfe am *Center* of my circling thought, / Onely *my selfe*

[71] Burton, *Anatomy of Melancholy,* 1, 4, and 12.
[72] Sandys, *Prudence,* 43 and 44. The square brackets are as the original.

I studie, learne, and *know.*" Likewise, Bishop Joseph Henshaw declared in
1637, "Account it the greatest knowledge truly to know thy selfe." For
Thomas Wright, self-knowledge leads to an understanding of "whither the
inclinations tend," that is, to knowledge of what kind of person one is. And
such knowledge allows one to take charge, to control the bodily urges; to become, indeed, fully human.[73]

Self-knowledge is, then, crucial to becoming a reasonable being. Bryskett
argued that "a man by knowing himselfe, becommeth in this life sage and
prudent, and vnderstandeth that he is made not to liue onely, as other creatures are, but also to liue well. For they that haue not this knowledge, are
like vnto bruite beasts."[74] The difference between the human and the animal is the difference between ignorance and self-knowledge, and that
knowledge has important implications for the capacity of the human.

Sir William Cornwallis wrote in 1610, "To know himselfe and the ap-
purtenaunces to himselfe is the vse of knowledge, and this knowledge vn-
maskes his eyes, & shews him wonders in himselfe, he becomes in this like
vnto God." The divine nature of the human is the outcome of self-knowledge: Cornwallis continues,

> To know himselfe, is to know before hand what may happen to himselfe, so
> shall he in despight of the apparitions of the world, stand vnmoueable: so
> shal he not be confined by expectation so shall hee not be seduced to thinke
> her ouerthrow his, but catch Poets description, and crowne himselfe with it;

> *Virtute præditi, & sapientis est viri,*
> *Non in rebus duris in Diuos fremere.*

> [It is the property of one endowed with virtue and wisdom not, in adversity,
> to bellow at the gods.][75]

Cornwallis is not uttering an overreaching blasphemy here: his man is "like
vnto God," not God himself, and the reason the simile works is that self-knowledge can allow man to be in control of himself and of time; to know his likely
failings and avoid them. Edmond Elviden put this in poetic terms in 1569:

> Al inconuenience that maye grow,
> or harmes that maye insue:

[73] Davies, *Nosce Teipsum,* 4 and 8; [Joseph Henshaw], *Meditations miscellaneous, Holy and Humane* (London: R. B., 1637), 53; [Wright], *Passions of the Minde,* 119.

[74] Bryskett, *Discovrse of Civill Life,* 166.

[75] Cornwallis, "Of Sorrow," in *Essayes,* sigs. Cc6ᵛ–Cc7ʳ. I am grateful to Barbara Goward and Erik Fudge for help with the translation.

Of diuers happs, the good forecaste
of wysdome may eschue.[76]

By knowing the past and applying it in the present, humans can plan their future and begin to overcome the chaos of the postlapsarian world. It should come as no surprise that an epigram included in a 1639 jest book had it that "Horse-keepers and ostlers (let the world go which way it will, though there be never so much alteration in times and persons) are still stable men."[77] It is by working with animals that the human becomes fully human, and stability is one outcome of prudence. The opposite of such planning and foresight is inconstancy, and Joseph Hall, the Bishop of Norwich, wrote in 1608 of this possibility:

> The inconstant man treads vpon a moouing earth, & keeps no pace. His proceedings are euer headdie and peremptory; for he hath not the patience to consult with reason, but determines meerly vpon fancie.[78]

To act with reason, then, is to act with self-awareness; is to know oneself. And the actions that follow such self-knowledge include understanding one's past, assessing one's present, and planning one's future. This is evidence of the chief among the virtues: prudence.

But being prudent is more than merely assessing the past and planning for the future. According to Bryskett, "there can be no vertue, where reason guideth not the mind. And for this cause wilde beasts (though they be terrible and fierce by nature) cannot be termed valiant, because they being stirred onely by naturall fiercenesse, wanting reason, do but follow their instinct, as do the Lions, Tygers, Beares, and such other like."[79] Once again, free will is crucial: an animal does not make choices; instead, it merely exists in a predisposed way, whereas a human uses reason, makes judgments, and acts on the basis of those judgments. For William Baldwin in 1564, "Vertue is none other thinge but dispositiō & exterior act of ȳ mīd, agreable to reason ād the moderation of nature."[80] As such, to be virtuous is to be reasonable, is to be human.

But being virtuous is not easy—it is not, like animal instinct, a predispo-

---

[76] Edmond Elviden, *The Closet of Counsells, conteining The aduice of diuers wyse Philosophers, touchinge sundry morall matters* (London: Thomas Colwell, 1569), fol. 1ʳ.

[77] *Conceits, Clinches, Flashes, and Whimzies* (1639), in *Shakespeare Jest-Books*, 3:36.

[78] Joseph Hall, *Characters of Vertues and Vices in two Bookes* (London: Melch. Bradwood, 1608), 107.

[79] Bryskett, *Discovrse of Civill Life*, 88.

[80] [William Baldwin], *A treatye of Moral Philosophy containing the sayinges of the wise* (London: Rycharde Tottill, 1564), fol. 90ʳ.

sition. Jewel writes that virtue "must be sought by a long and a continuall exercise of commanding your selfe, bridling your affectiōs, & pursuing good without intermission; forasmuch as discontinuance begetteth vice." He continues: "None intitle themselues prudent, valiant, iust, or by the names of any other vertues, but such as perseuer with immoueable constancy."[81] And constancy, in Caussin's terms, "is the type, and Crowne of prudence."[82] But this call for self-command and its ensuing constancy is not a kind of Stoicism. Jewel writes, "vertue teacheth vs, not to depriue our selues of wishes and desires; but that we should continually represse and master them."[83] This is very different from the ideas put forward by early modern Stoics. Guillaume Du Vair, for example, whose 1585 *Philosophie morale des Stoïques* was translated into English in 1598, writes that "to live according unto nature is, not to bee troubled with any passions or perturbations of the minde": true happiness, in his opinion, lies in the purging from the mind of all passions.[84] As Anthony Levi has noted, for the Stoics, whose work was undergoing a revival in the sixteenth and seventeenth centuries, "passions were false judgements." Apathy, in their view, meant "absence of passion, [and] was simply the equivalent of rationality."[85] This view was not accepted by all early modern writers. In fact, Michael C. Schoenfeldt has argued that the debate "between the eradication of emotion that constitutes the Stoic ideal of *apatheia* and the proper direction of emotion that comprises Christian devotion" was "one of the central philosophical struggles of Renaissance culture."[86] The key criticism of the Stoic position was that Christ experienced passions and that therefore, in Wright's words, "it were blasphemous to say, that absolutely all passions were ill."[87] Likewise, Joseph Hall wrote in 1607, "I would not bee a Stoïcke, to haue no Passions: for that were to ouerthrowe this inward gouernment God hath erected in me; but a Christian, to order those I haue."[88]

So, for these thinkers, humans can reveal the power of their reason through their self-knowledge; can reveal themselves to be good Christians, living not in a world without temptation but with the possibility of seduction

[81] Jewel, *Golden Cabinet*, 10–11.

[82] Caussin, *Holy Covrt*, 352.

[83] Jewel, *Golden Cabinet*, 130.

[84] Guillaume Du Vair, *The Moral Philosophy of the Stoicks* (London: Felix Kingston, 1598), 30.

[85] Anthony Levi, S.J., *French Moralists: The Theory of the Passions 1585–1649* (Oxford: Clarendon, 1964), 4 and 31.

[86] Michael C. Schoenfeldt, *Bodies and Selves in Early Modern England: Physiology and Inwardness in Spenser, Shakespeare, Herbert, and Milton* (Cambridge: Cambridge University Press, 1999), 85.

[87] [Wright], *Passions of the Minde*, 28.

[88] Joseph Hall, *Meditations and Vowes Diuine and Morall* (London: Humfrey Lownes, 1607), Book I, 98–99.

by the passions, which they resist. Animals, on the other hand, have no such choices, and it is the decision to refuse the passions that reveals the reasonable being. This is not to return to the equilibrium of body and mind of the time before the Fall, rather it is to engage reason in a struggle with the body and win. And the struggle is crucial.

It is in this way that Cornwallis's proposal that humans become "like vnto God" is comprehensible: humans cannot become God, but they can attempt to replicate on an earthly level what God is capable of on a universal one. As Sandys wrote, "The Providence of God is the certaine knowledge of God, conceived in his Vnderstanding from eternity, concerning those things which were necessarily and contingently done in times past, or which any time are to come."[89] God's power manifests itself in His capacity to transcend time: because He knows the past, the present, and the future, the Almighty will not be altered as time passes. By careful acts of recollection, and insightful analysis of possible futures—by being, in short, prudent—humans (only humans) can achieve something similar.

The Dream of the Human

Sometimes the future can be foreseen, however, not through planning but through inspiration. Sometimes reason seems to operate outside of the human. In *A Treatise of Specters* (1605), a summary and translation of the French original, Pierre Le Loyer repeats the orthodox arguments outlined in this chapter so far: "there are," he writes,

> two sortes of Imagination, namely, one Intellectuall, and without corporall substance: The other sensible and corporall. The Intellectuall is the Fantasie, of which is bred and engendred in vs a memory or remembrance (as the *Peripatetickes* speake) and the discourse of the reasonable soule: I meane that discourse which is proper only vnto man: by the which he ballanceth and weigheth the things present, by those which are past, & foreseeth by things past, those which are to come after.

Le Loyer goes on, however, to draw a link with another property of the human. Dreaming, he writes, is equivalent to "discourse"; it is an aspect of reason.[90] This is because, for some early modern writers, true dreams were believed to give the dreamer the opportunity to see beyond the present and

---

[89] Sandys, *Prudence*, 103.

[90] [Pierre Le Loyer], *A Treatise of Specters or straunge Sights, Visions and Apparitions appearing sensibly vnto men* (London: no publisher, 1605), sig. B3ᵛ.

the material. Writing in 1628, the moralist Owen Feltham regarded dreams as a "notable *meanes* of discouering our owne *inclinations.*"[91] That is, they fulfill the function that Wright found in self-knowledge. They give the individual the possibility of a fuller awareness of the self that can in turn be directed toward living a virtuous life. Sir Thomas Browne agreed with Feltham's assessment, writing that through dreams "wee may more sensibly understand ourselves."[92]

But dreams were also understood to have another function: they not only reveal the inclinations of the dreamer but are also potentially prophetic. In the fullest examination of the meaning and power of dreams in early modern England—*The Moste pleasaunte Arte of the Interpretacion of Dreames* (1576)—Thomas Hill proposed that dreams could offer "cōfort"; that they could, by foretelling future events, enable the dreamer to "labour to preuent or hinder the imminent mysfortune, or at the least arme himselfe so stronglye wyth patience as quietly to beare them: for a mischiefe knowen of before, and diligētly loked for, is not so greuous as whē it commeth on a sodayne."[93] To be able to use this faculty, however, humans must learn the art of divination, and one of the fundamental issues that must be learned is which dreams are worthy of interpretation; which dreams are prophetic.

Echoing classical and medieval ideas, Hill proposed that there are two kinds of dreams available to humans: vain and true. Vain dreams, he writes, are "no true signifiers of matters to come but rather shewers of the present affections and desiers of the body." True dreams, on the other hand, "signifie matters to come." The distinction here is between dreams that are experienced by those "ouercharged with the burthē of meate or drinckes, or superfluous humors," and those "seene by graue & sober persons"; between those dictated by the body's excesses and those by the mind whose body is held in abeyance.[94] A vain dream in these terms is merely reactive—it relies on the existence of the material—whereas a true dream is of the rational soul and is prophetic: it can envision that most immaterial thing, the future.

This categorization was extended thirty-one years later by Thomas Walkington. Whereas Hill proposed one category of vain dreams, Walkington has two: a vain dream, for Walkington, is "whē a man imagins hee doth such things in his sleepe, which he did the day before: the species being strongly fixed in his phantasie." This is different from what he terms a "dreame Nat-

---

[91] Owen Feltham, *Resolves or, Excogitations. A Second Century* (London: Henry Seile, 1628), 152.

[92] Thomas Browne, *On Dreams* (n.d.) in *Sir Thomas Browne: The Major Works,* edited by C. A. Patrides (London: Penguin, 1977), 477.

[93] Thomas Hill, "Preface to the Reader," in *The Moste pleasaunte Arte of the Interpretacion of Dreames* (London: T. Marsh, 1576), n.p.

[94] Hill, *Moste pleasaunte Arte,* "Epistle Dedicatory," n.p.

urall" which "ariseth from our complections, when humours beene too aboundant in a wight."[95] Walkington has recognized that there are two ways in which a human can dream without the use of the rational soul: through the overwhelming of reason by the body, here figured in the overabundance of humors manifesting themselves during sleep, and by living only in the present, that is, by experiencing only the most recent events of life in dreams. Robert Burton wrote, "As a dogge dreames of an hare, so doe men, on such subiects, they thought on last."[96]

The fact that animals experienced dreams was not doubted in the early modern period: the natural philosopher and divine Edward Topsell wrote of dogs in 1607, "They sleepe as doth a man, and therein dreame verie often, as may appeare by their often barking in their sleepe."[97] And the distinction between true and vain (and natural) dreams allows for this fact. The dreams of animals are not to be accounted true—prophetic—dreams. Rather, the dreams that animals experience are of the body, and deal only with those events that remain imprinted on the brain or are the result of indigestion. As Le Loyer put it:

> For albeit the vnreasonable creatures doe sometimes seeme to haue a kind of discourse, or dreaming in them, (as is to be seene in Horses and Dogges) yet this dreaming or discourse in them, is no other, then meerely bestial and brutish: which doth not accomodate nor apply it selfe, but onely to things present, by an vnreasonable appetite & desire vnto those things which they loue, and by eschewing and abhorring to their vtmost powers, that which may be fearefull or contrary vnto them.[98]

It could not be otherwise because animals, of course, do not have a rational soul that might offer anything else.

So, a true dream is another evidence of reason, something that is made clear by Sir Thomas Browne, whose distinction between types of dream upholds the meaning of previous models but changes the terminology in a way that is inherently logical. For Browne the distinction between true and vain becomes the distinction between "Angelicall" and "animal" dreams,[99] and humans, as Sir Walter Raleigh noted, are beings caught between these categories. As was written in the anonymous "Epistle Dedicatory" to the 1606 translation of Artemidorus's *The Iudgement, or exposition of Dreames*:

[95] T. W. [Thomas Walkington], *The Optick Glasse of Hvmors* (London: John Windet, 1607), sigs. 76[v] and 77[r].

[96] Burton, *Anatomy of Melancholy*, 238.

[97] Edward Topsell, *The Historie of Fovre-Footed Beastes* (London: William Jaggard, 1607), 139.

[98] [Le Loyer], *Treatise of Specters*, sigs. B3[v]–B4[r].

[99] Browne, *On Dreams*, 475 and 476.

what is ther more honest, more holy, & that comes neerer to Diuinitie, then that a humane spirit may support, know, and vnderstand a part of future things without offending God, only by the meanes of auses [auspices] & significations præcedēt, which are sent to vs by him, which is onely proper to man, who hath alone the vse of reason, whereby he may discerne, and iudge of things to come.[100]

But, despite all these claims for the prophetic power of a true dream, what is still unclear is how to tell if a dream is true or vain; how to tell if it is a vision of the mind or of the body. Whereas a true laugh can be judged to be true—and not dog or bastard—by the observer's analysis of the object of the laugh, it is impossible, on the basis of the dream itself, to make such a judgment. Apart from those that merely repeat the events of the day (and so can clearly be dismissed as vain), dreams cannot in and of themselves be judged to be either true or vain, since that which is depicted in them has not yet, and may never, happen. The status of a dream as true can only be asserted once the event that is apparently prophesied has taken place, and that is not something that is to be judged in the evidence given by the dream itself, but rather by the events that follow the dream. In itself the dream—unlike the laugh—is meaningless. Moreover, as Aristotle noted of dreams in *De divinatione per somnum*, "it is not improbable that some of the presentations which come before the mind in sleep may even be causes of the actions cognate to each of them."[101] This idea is repeated by the German writer and magician Henry Cornelius Agrippa in the sixteenth century: diviners of dreams, he writes, "searche out hidden things either by aduenturous chaunce, or by the mouing of the minde."[102] That is, prophetic dreams prophesy merely by luck or because they themselves make the things prophesied happen rather than simply tell that those things will happen anyway. True dreams, in fact, can be self-fulfilling prophecies, and self-fulfilling prophecies are not the same as prophecies. Thus what is true and what vain, reasonable and unreasonable, human and animal, becomes difficult to assess. The human dream, in fact, becomes a dream of the category of the human in which the human is a deferred presence rather than an innate one.

But a solution was offered to this conundrum. The early modern translator of Artemidorus offered this possibility for discriminating between types of dreams:

---

[100] "The Epistle Dedicatory," in Artemidorus, *The Iudgement, or exposition of Dreames* (London: William Jones, 1606), n.p.

[101] Aristotle, *De divinatione per somnum* 463a, translated by J. I. Beare, in *Works of Aristotle*, 1:707.

[102] Henry Cornelius Agrippa, *Of the Vanitie and vncertaintie of Artes and Sciences* (London: Henrie Bynneman, 1575), sig. 50ʳ.

And surely I belieue that an ordinary Whooremaister, vsurer, enuious, ambitious or dronkenman shall not commonlye see any dreame which con-
cernes his honour, the good or profit of him or his, or common wealth. But
a good, vertuous pure, and cleane, (because he is exempt from humaine
fragility) I thinke may and shall often see & interpret dreams and visions, to
the safety, honour, and profit of himselfe, his friends, and the common
wealth, for as much as his spirite is lesse bound, tyed and soyeld with the fel-
lowshippe of the body.[103]

Here is a link between virtue and the human that we have come across before. Only the "good, vertuous pure, and cleane" man can experience true
dreams because only the "good, vertuous pure, and cleane" man is truly exercising reason, and therefore is in possession of that which is the source of
a true dream. This conceptualization of dreams helps to explain Owen
Feltham's claim that dreams are "the naked and naturall thoughts of our
soules."[104] If humans use only their sensitive soul, then their dreams likewise
will be sensitive; that is, of the body. A sensual man will have vain dreams,
and, as Hill wrote, "graue & sober persons" will have true ones.

Thus we have what might be termed a virtuous circle: a good man dreams
true dreams, and the interpretation of those true dreams enables the good
man to maintain stability in a changing world; to be, in fact, prudent. What
emerges is a clear link between virtue and reason; between being good and
being rational. But, being a human is not a simple thing. The being called
human, by failing to use its rational part, risks losing its humanity. Such failures are almost always figured as a descent to the level of the animal. A bodily laugh is a dog laugh; an animal can never be virtuous; a horse does not
know it is a horse; and a dog can only dream of the hare it was chasing that
same day. These are the limitations that the discourse of reason places on
animals, and these are the limitations that support the apparently limitless
capacity of the human.

This is a point made with a particular clarity in the 1606 translation of
Artemidorus's dream text, where, under the heading "Of beasts of all sorts,"
the meanings of animals in dreams are outlined: a sheep signifies, for example, advancement, "wherefore it is good to dreame you haue many of thē,
or see them of others and feed them"; an ass signifies a wise companion; "To
see a gentle, familiar and fawning Lyon, signifies good, and profit," and so
on. Artemidorus then goes on to note that in dreams animals "signifie exceeding profit if they speake our lāguage, especially if they say any good
thing or joyfull; and all which they speake commonly falls out."[105] For

---

[103] "Epistle Dedicatory," in Artemidorus, *Iudgement*, sig. B1r–v.
[104] Feltham, *Resolves*, 152.
[105] Artemidorus, *Iudgement*, 73.

Artemidorus, the speaking animal of the dream is the vision of the ideal because, by his account, what an animal says "commonly falls out." That is, in the true dream of the speaking animal the presence of the human is no longer postponed but is confirmed within the dream itself because the dream, by its nature, will come true.

It is not, I think, too big a leap to see how this interpretation of the speaking animal in the dream offers a vision of the way in which animals are used more generally in the discourse of reason in the early modern period. Animals "speak" to humans in the texts written by humans that confirm their own rational status, and as such these "speaking" animals—the horse that does not know its way home until it is on the well-trodden path, the dog that cannot tell a human from an animal—confirm the superiority of humans. It is interesting to wonder how the discourse of reason would have worked without animals. If rationality has a relative status—opposed always to irrationality—it would be impossible to make a judgment about the possession of reason without invoking the potential for its lack. On this basis, just as the speaking animal in the dream proves the human to be truly human by "commonly" speaking the truth, so the human, in the wider discourse, is reliant on animals for its status. In a world without animals, humans would not only lose companions, workers, sources of food, clothing, and so on; they would lose themselves.

Empirical Proofs

But as well as—or perhaps because of—the internal and eternal struggle for reason's dominion, and the necessity for animals in the definition of the human, another problem faces the discourse of reason. What is not clarified in any of the statements about human difference is how, knowing that a human has access to the past and the future—to change, self-knowledge, abstraction, and so on—a human can be found. Or, to put it another way, these writers know conceptually what a human is, and how it differs from an animal, but they have not fully explained how they would be able to know one when they saw one. In this way, searching for the true human is like searching for the true dream.

Raleigh noted this difficulty: man, he wrote, "is ignorant of the Essence of his owne soule."[106] This would seem to be something of a dilemma. If the foundation of all difference, all hierarchy, is unknown, can that hierarchy persist? The answer, of course, is yes. Writing in 1637, Joseph Henshaw, for instance, had a simple solution: "The goodnesse of the minde, is witnessd in

---

[106] Raleigh, *Historie of the World,* sig. B5ᵛ.

the outward actions; the goodnesse of the outward actions, is determined by the intention and minde."[107] Being and acting are inseparable; one reveals the other. Raleigh likewise sees action as the best reporter of mind: "And though it hath pleased God to reserue the Art of reading mens thoughts to himselfe: yet, as the fruit tels the name of the Tree; so do the outward works of men (so far as their cogitatiõs are acted) giue vs wherof to gheß at the rest."

But if, as Raleigh notes, outward works allow evidence from which only *guesses* can be made, then judgment of displays of reason, and therefore of human status, cannot be absolute. There is not only the passage of thought from mind to action to assess; there is also the parallel passage in the mind of the judge from the exterior, observed action to the mind where judgment takes place. (And the judge can, of course, be the self performing the action.) In these translations of the immaterial to the physical, the physical to the immaterial, errors can occur, *mis*judgments can take place. Even among the "better sort," Raleigh writes, "euery vnderstanding hath a peculiar iudgement, by which it both censureth other men, & valueth it self." He continues, "euery one is touched most, with that which most neerely seemeth to touch his owne priuate; Or otherwise best suteth with his apprehension."[108] In fact, there emerges a difference between judgment and mere opinion, and, for Sir William Cornwallis, this distinction concerns the use of reason: "Opinion . . . is the destinated censure of Affection, as iudgement is the soules."[109]

Because of these dangers—this capacity to judge only on the basis of one's own desires, and the application of affection rather than reason—there is a crucial difference between human and divine wisdom. Raleigh writes, "the iudgements of GOD are for euer vnchangeable; neyther is hee wearied by the long processe of time, and won to giue his bleßing in one age, to that which he hath cursed in another."[110] Humans' judgments, on the other hand, are changeable, impermanent, and so free will, which had been represented as a crucial human capacity, becomes a limitation. As such—because reason is to be judged in action, not in mere possession (for who can see the invisible?)—the judgment of reason can itself turn out to be unreasonable.

But we can trace this unraveling of logic on a wider level as well. We have shifted from what was an apparently clear statement of fact—humans are reasonable, animals unreasonable—to a discussion of something far more

---

[107] [Henshaw], *Meditations miscellaneous,* 50.
[108] Raleigh, *Historie of the World,* sig. A^v–A2^r.
[109] Cornwallis, "Of Affection," in *Essayes,* sigs. N6r–^{r}–^{v}.
[110] Raleigh, *Historie of the World,* sig. A2^v

ambiguous, subjective: the judgment of the actions of oneself and others. To be judged by one's actions means not only that different assessments of the same action might occur but also, and perhaps even more threatening, that a human could perform what might logically be an inhuman action; that a human could apparently sink to the level of an animal by failing to use that essence that places humans above animals. What does it mean to get drunk, or to be cruel, for instance? How can these actions be assessed in terms of their relation to reason, and therefore to human immortality?—for this is how they have to be assessed. Or, to take a step back, how can an infant, or a child, be said to be human when it displays none of the properties of humanity? Is being human a process—does one become human—or are learning and developing parts of being (not becoming) human? The intangibility of the difference between humans and animals—the invisibility of the rational soul—means that all that can be relied on is empirical evidence; is what those beings that are called human actually do. And sometimes it is difficult to see the rational soul in action, and therefore difficult to see the thing that makes the human human.

Early modern writers began to differentiate humans and animals from a solid foundation, then, but swiftly found themselves in a poorly built, fragile building. They began with absolute difference and slowly discovered that that absolute is compromised, is itself unreasonable. But, of course, the emergence of this compromised absolute does not lead to its abandonment; the logical breakdown does not affect the persistence of the logic. The foundation of the discourse of reason—that humans and animals are crucially different on the basis of their possession or lack of reason—remains even as it is challenged by the very logic that proclaims it, and it is to the challenges that this book now turns.

CHAPTER 2

Becoming Human

When does one become a human? Or rather, when is human status conferred on the human? If, as it was in the early modern period, human status is designated on the basis of the possession of the inorganic, rational soul, when does that soul come into being? Is the human—even in embryo—to be considered always human? Is the human human from the moment of birth? Or is the gaining of the rational soul a process at the end of which a human being emerges? The answer to this latter question is of great importance here, for if the human exists in utero, then that might offer a confirmation of difference; it might make absolute and natural the distinction between human and animal. But if the human is revealed to be created after birth, who could say whether the human is *always* fashioned, whether the process of ex utero creation always works? In attempting to answer these questions I am looking at the origin of the difference between humans and animals, and it is an origin, I will argue, that is as confused as many of the other issues concerning the possession of reason already outlined.

Limited Designations

In the early modern texts already cited the focus is, apparently, on the human as a general, catchall category. What is clear, however, is that, in many of these texts, this human is actually a particular kind of being. The human is usually referred to with the masculine pronoun "he" and with "man" as an apparent synonym for "human." This could be interpreted as being merely the style of early modern writers, but even so this style does give a priority to male humans and in many ways removes female humans from the debate.

The priority of the male in early modern England has been emphasized in much recent scholarship, but a complete exclusion of women from debates about reason cannot exist for very simple reasons. Any distinction between male and female which excluded women from the possession of the rational soul would have made the status of the infant—created by the mingling of male and female fluids, born from the woman's body—worryingly hybrid. It would also have meant that a man—the possessor of a rational soul—could play a part in the reproduction of an irrational being: a daughter. This would have been a biological nightmare, but it did not preclude discussions denying the reasonable (and immortal) status of women. In 1595, for example, "an anonymous German humanist" published a treatise entitled *A New Discourse Against Women, In Which It Is Proven That They Are Not Human.* This text, a parody, as Bruce Boehrer has noted, of "the scriptural literalism popular with radical reformed Christianity in general," makes the case that, according to an analysis of scripture, women "do not have souls and therefore are incapable of salvation."[1] This proposal of absolute difference is not reiterated in the period, but what is claimed is that women have a lower status because of a natural deficiency. Juan Huarte, whose *Examen de Ingenios* was translated into English in 1594, claimed that because God originally gave Eve less reason than Adam (hence the Devil tempted the woman, not the man), "*the naturall composition which the woman hath in her braine, is not capable of much wit, nor much wisedome.*"[2] A very simple analogy was formed from this idea: in William Strode's *The Floating Island,* a play commissioned by Archbishop Laud and performed before Charles I in 1636, a civil war is waged by the passions against King Prudentius. Strode wrote, "since by Passion this revolt is made / From Reason unto Sense, the Rule should passe / From man to Woman."[3] The sensuality of women was understood to restrict their capacity for virtue. For example, John Knox, with exquisitely bad timing, wrote in 1558 (the year Elizabeth I came to the English throne), "Nature I say, doth paynt [women] further to be weake, fraile, impacient, feble and foolishe: and experience hath declared them to be unconstant, variable, cruell and lacking the spirit of counsel and regiment."[4] This natural weakness was the cause of a natural (that is, unthought) clarity of emotion: according to Thomas Wright, women's passions "may easely be discouered"

---

[1] Bruce Boehrer, *Shakespeare among the Animals: Nature and Society in the Drama of Early Modern England* (New York: Palgrave, 2002), 58–59.

[2] Juan Huarte, *Examen de Ingenios. The Examination of mens Wits* (London: Adam Islip, 1596), "Epistle to the Reader," n.p.

[3] William Strode, *The Floating Island: A Tragicomedy, Acted before his Majesty at OXFORD, Aug.29.1636* (London: H. Twiford, 1655), I Scena v.

[4] John Knox, *The First Blast of the Trumpet against the Monstrous Regiment of Women* (1558), in *Lay By Your Needles Ladies, Take the Pen,* ed. Suzanne Trill, Kate Chedgzoy, and Melanie Osborne (London: Arnold, 1997), 33.

since women are not able to hide or subdue them as men are. Wright argued that this is why "women, cannot abide to looke in their fathers, masters, or betters faces, because, euen nature it selfe seemeth to teach them, that thorow their eyes they see their hearts."[5] Women's perceived inability to counter the urges of the body, evidence of the weakness of will, was repeated by Nicholas Coeffeteau, who wrote: "generous spirits are as it were, impenetrable to offences; whereas they that cannot resist, shew their weakenesse; whereby we see that women, children, sicke folkes, and olde men are most subiect to these motions and impressions."[6] The external world impacts on women, and on what we might now term the other vulnerable human groups, in a way that implies that their control of the internal world of reason is frail. In these terms, women were certainly human, but their humanity was perceived to be more fragile, and as such somehow closer to— although always different from—animals. And this physiological and attendant rational differentiation of male from female naturally supports, among other things, an economic system that precludes most women from property ownership.

However, the potential for a distinction between the sexes in discussions of reason is not the only limitation of the apparently catchall title "human." Most important in this chapter is the fact that in most early modern discussions, the human is represented as an adult; as a being who has, by implication, received an education—however rudimentary—and knows (or should know) right from wrong. The status "reasonable," in fact, is designated to include only those humans who are capable of engaging with the debate itself. As John Davies wrote,

> And as if beasts conceiud what Reason were,
> And that conception should distinctly show;
> They should the name of *reasonable* beare,
> For without *Reason* none could *reason* know.[7]

But what would happen to these discussions, so focused as they are on the male adult, if they turned to think about the embryo, the infant, or the child? If the embryo, the infant, and the child are said to possess a rational soul, but they fail to display their possession of this soul, how can they be judged to be reasonable; that is, human? Such questions are inquiring into the nature of the human, into the nature of the emergence of the human in childhood and before, and into early modern understandings of the is-

---

[5] [Thomas Wright], *The Passions of the Minde* (London: V. S., 1601), 54.

[6] Nicholas Coeffeteau, *A Table of Humane Passions. With their Causes and Effects* (London: Nicholas Okes, 1621), 620.

[7] John Davies, *Nosce Teipsum* (London: Richard Field, 1599), 65.

sues at stake in the operations of the rational soul. If the human is a natural—given—category, then the moment when the rational soul comes into
being is crucial to understanding the character of that natural status. If this
is not the case—if reason cannot be displayed until something that is not
natural but cultural has taken place—this too should alert us to some important questions about the nature of the human. I will start at the very beginning, the moment of the creation of the embryo; the human will mature
as this chapter progresses.

Born and Made

In the early modern period it was believed that the creation of a human began with the mingling of the parents' seminal fluids. In a text first published
in 1548 Thomas Vicary proclaimed, the "Embreon is a thing ingendred in
the mothers wombe, the original wherof is ŷ sparme of the man and the
woman, of the which is made by the might and power of God, in the mothers wombe a chylde." This act of generation is wholly natural and clearly gendered: "in man," Vicary writes, sperm "is hotte, white, & thicke," whereas,
"the womans sparme is thinner, colder, and feebler." And the act does not
involve the rational soul. Indeed, Vicary uses an analogy, first found in Aristotle's *De generatione animalium*, that reinforces the fact that reproduction is
the work of the vegetative soul: "lyke as the Renet of the Cheese hath by him
selfe the way or vertue of working, so hath the mylke by waye of suffering:
and as the Renet and mylke make the cheese, so doth the sparme of man
and woman make the generation of Embreon."[8] Here the woman is the passive partner in the sexual relationship, although she remains human.

What follows the sexual act is pregnancy, and the confirmation of pregnancy comes, said James Guillemeau in 1612, "when he begins to stirre and
mooue" in the womb. The embryo is gendered here, and this gendering
continues into medical fact: Guillemeau proposes, again following convention, that "the male child beginneth to stirre at the end of the third moneth, or sooner; and the female at the third or fourth moneth."[9] By this point
in the pregnancy the embryo has become a human, in that it has developed
both body and, importantly, soul.

The development of the body of the embryo presented by Vicary and
other early modern commentators broadly follows the order described by
Aristotle and Galen. Vicary writes:

[8] Thomas Vicary, *A profitable Treatise of the Anatomie of mans body* (1548; repr., London:
Henry Bamforde, 1577), sigs. M3r–M4r.

[9] James Guillemeau, *Child-Birth Or, The Happy Deliverie of Women* (London: A. Hatfield,
1612), 6 and 16.

the fyrst thing that is shapen be the principals, as is the Harte, Lyuer, and Brayne. For of the Hart springeth the Arteirs, of the Lyuer the Veynes, and of the Brayne the Nerues: and when these are made, Nature maketh & shapeth Bones and grystles to keepe & saue them, as the bones of the head for the Brayne, the Brest bones and the Ribbes for the Harte and the Lyuer. And after these springeth al other member one after another.

He also proposes that there are four stages in the development of the embryo; three before it possesses a soul and becomes human.

The first is, when the sayde sparme or seede is at the fyrst as it were mylke: The seconde is, when it is turned from that kinde into another kinde, is yet but as a lumpe of blood, and this is called of Ypocras, *Fettus:* The third degree is, when the principals be shapen, as the Hart, lyuer, and Brayne: The fourth and laste, as when al the other members be perfectly shapen, then it receyueth the soule wyth life and breath, and then it beginneth to moue it selfe alone.

These stages of development allowed Vicary and other thinkers to observe a timescale in the development of the embryo from sperm to human. According to Vicary's sums, there are "xlvj.dayes from the day of conception vnto the day of ful perfection and receyuing of the soule, as God best knoweth."[10] In writing this, Vicary is following Aristotle's proposal in *Historia animalium* that the male child is formed after forty days. What Vicary does not comment on is Aristotle's suggestion in the same place about the female child.[11] In this regard, John Woolton seems a more faithful Aristotelian, proposing that the "childe is in bodye parfyted in his mothers wombe . . . in a manne childe . . . aboute the fourty daye, in a woman childe aboute the 90 day."[12] The speed of the full emergence of the male is linked to the preponderance of female sperm in the making of a female child, and of the relative weakness of that sperm. What is also linked here is perceived strength of the masculine sperm and the masculine will.

It is at the point of the perfection of the body that the embryo begins to move in the womb, and this movement signifies the status of the child. As William Hill noted in 1605, "Nothing can bee, without that from which it receaueth his beeing: nothing can mooue, without a moouer: nothing can be fully formed, without *forma*. But the Soule is the first moouer, the first acte;

---

[10] Vicary, *Anatomie of mans body,* sigs. M4r–v.

[11] Aristotle, *Historia animalium* 583b.

[12] John Woolton, *A Treatise of the Immortalitie of the Soule* (London: John Shepperd, 1576), fols. 18r–v.

it is the life of the Body, and it is *forma hominis,* the forme of Man."[13] Because of this priority of the soul—it is the first mover in the human, as God is the first mover in the universe—so the infant must be in possession of the rational, immortal soul before it can move, and therefore must be in possession of it before its birth.

The infant's in utero possession of the rational soul has important implications for its status. On one level, as Pamela M. Huby has noted, "The embryo is animal before it is human."[14] That is, the animal capacities of the sensitive soul are created before the rational soul appears. But, alongside this is the belief that no human can be born without a rational soul, and therefore any being that lacks a rational soul at birth cannot be human.[15] But possession of the rational soul does not mean, as we have seen in the case of laughter, that that soul gains automatic priority. The Fall inevitably undid that possibility, even before birth. William Hill wrote that "the wombe is the prison in which the Soule is imprisoned."[16] John Davies likewise regarded the corruption of the soul as being a result of the Fall, and agreed with Hill's assessment of the lack of innocence of even the prenatal soul:

> And yet this Soule (made good by God at first,
>   And not corrupted by the Bodies ill)
>   Euen in the wombe is sinfull, and accurst,
>   Ere she can Iudge by wit, or choose by will.[17]

From this Calvinist perspective (Calvin argued that infants were "originally depraved")[18] what was immortal and rational had to strive against what was mortal and sensitive prior even to birth, and there exists a tale of one infant who clearly managed to overwhelm his sensual passions with his precocious rational will. In 1599 Anna Iacobs and her husband John Martinson heard their unborn infant crying in its mother's womb and were both "heereat greatly amazed." Soon afterwards the child was born but became ill, and in the evening of 14 January, fifteen days after its birth, the infant cast his eyes up to heaven, thrust out his arms and fists, and

---

[13] William Hill, *The Infancie of the Soule: Or, The Soule of an Infant* (London: C. Knight, 1605), sigs. C^v–C2^r.

[14] Pamela M. Huby, "Soul, Life, Sense, Intellect: Some Thirteenth-Century Problems," in *The Human Embryo: Aristotle and the Arabic and European Traditions,* ed. G. R. Dunstan (Exeter: Exeter University Press, 1990), 118.

[15] There is a debate concerning the origin of the soul—whether it is created by God or by the mingling of the parents' sperm. This debate does not challenge the existence of the in utero possession of the rational soul, and so is not dealt with here, but John Woolton offers an overview of the various positions in this argument. Woolton, *Immortalitie of the Soule,* fols. 19^r, 26^v–27^r, and 27^{r-v}.

[16] Hill, *Infancie of the Soule,* sig. B4^v.

[17] Davies, *Nosce Teipsum,* 31.

[18] John Calvin, *Institutes of the Christian Religion,* trans. Henry Beveridge (London: James Clarke, 1949), 1:215.

(through anguish) spake these words three times.

O my God.

O my God.

O my God.

Which words were spoken with such distinction, that betweene each time one might haue spoken two or three words: also they seemed not to be the words of a sick and feeble infant, but as if they had beene of a childe of ten yeares of age: and more ouer so lowde, that they might haue beene heard to the Church.

Later the child became weaker and was heard to say the words "Aye me" "how often it is not remembred." The moral of this extraordinary tale is clear when the infant's father invokes the "viii. Psalme, *That Gods might and power is disclosed by the mouthes of sucking Babes.* By this," the pamphlet notes, "may be seene, that the little Babes also haue the holy Ghost."[19] To have the Holy Ghost must mean that the child has a rational (immortal) soul with which his body will be reunited after Judgment Day.

But, of course, this tale of a speaking infant is interesting because it is "strange." Most newborn babies cannot give displays of reason or faith such as this, and so they cannot be judged to be reasonable creatures. The infant has no language with which to express thoughts, has almost no capacity for self-willed movement, and has no capacity to judge the world around it because everything is new, and therefore everything is equally significant. How then can this being be assumed to be human?

Physiology was sometimes offered to explain the infant's apparent lack of reason. William Hill argued that "This reasonable Soule at first genera-tion of Man, is plunged and infused *Multo humore a quo vires offenduntur, cali-gantur et obtenibrantur:* with much moysture, from the which the powers be hurt, blinded, and made darke." The moistness of the body means that "In-fants seeme to be voyde of reason at the first birth." Hill goes on to argue that this physical weakness is not permanent, and does not affect the nature of the soul itself:

But those humors in time deminishing, and the Body being made more dry, it sheweth further power. This caused one to say, That if the Seede, & Men-struall blood, which be the two materiall principles of which we be fashioned, were cold and dry, as they be hot & moyst, that Children should be able for to reason.[20]

[19] *A Strange and Miraculovs Accident happened in the Cittie of Purmerent, on New-yeeres euen last past 1599* (London: John Wolfe, 1599), 5–6 and 7.

[20] Hill, *Infancie of the Soule*, sig. C4ʳ.

In this interpretation, it is not the incapacity of the rational soul that pre-cludes infant reason; it is the incapacity of the body into which the soul is placed. So if the human is a being in possession of a rational soul, then clearly the infant is human; it just has not gained the physical capacity to express its human-ness at this stage in its development. As Edward Jorden wrote in 1603: "nature brings vs not forth into the world perfect men." He continues, arguing that it is time—"when wee come to full growth"—that can give "that ripenesse and integritie, of all humaine actions which afterwards we attaine vnto."[21] Similarly, Lodowick Bryskett went so far as to draw an analogy between the lowest kind of soul and the status of the child:

> As the soule of life therfore, called *vegetatiue*, is the foundatiō of the rest, and consequently of the basest: so is the age of childhood the foundatiō of the other ages, and therfore the least noble, for the necessity which it carieth with it.[22]

By implication, infant is to plant what adult is to human; or, in Seneca's terms, "no more is an infant capable of good as yet, then a tree, or any dumbe beast. But why is not good in a tree or dumbe beast? Because reason is not in them, and therefore is it not in an infant, by reason that he wanteth reason, whereunto he hath attained, he shall approch more goodnessse."[23] Bryskett writes, "being yet but new in the world, and not acquainted with those things, the images whereof are presented to them by the senses of hearing and seeing; they easily giue themselues to way-wardnes and crying, when they see any strange sight or images, or heare a fearefull sound or noise." But this stage in the development of this being called human is not the worst; that is yet to come. It is worth quoting Bryskett at length here, as he singles out youth for this honor, using a familiar image:

> And as it is harder to rule two horses to guide a coach or charret then one: so is there farre greater difficultie in guiding a yong man then a child: for he is stirred much more with passions then the simple age of a child, and is more violently caried away with things that delight him; because he hath now the second power of the soule in force to draw him, which for the most part is much more contrary to reason then the first. For wheras that first coueted only that which was profitable, and which might nourish the bodie without any great regard of that which was honest, as whereof it had no knowledge

[21] Edward Jorden, *A Briefe Discovrse of a Disease Called the Suffocation of the Mother* (London: Iohn Windet, 1603), sig. 17ᵛ.

[22] Lodowick Bryskett, *A Discovrse of Civill Life* (London: Edward Blount, 1606), 45.

[23] Seneca, "Epistle CXXIIII," in *The Workes of Lvcivs Annævs Seneca, Both Morall and Naturall*, trans. Tho. Lodge (London: William Stansby, 1614), 491

at all; this other being wholy bent to delight, respecteth little any other thing: which delight hauing greatest force in yong mindes, draweth them sundry wayes, and by allurements maketh them so much the more greedy to attaine the things they take pleasure in, as the spurres wherewith they be pricked are more sharpe and poignant.

It is for this reason that Bryskett can write, with no sense of irony, that "For though man be framed by nature mild and gentle, yet if he be not from the beginning diligently instructed and taught, he becometh of human and be-nigne that he was, more fierce and cruell then the most wild and sauage beast of the field."[24]

This sense of the dangerous potential of the human and, by implication, the perception that the rational soul may fail to gain ascendancy even as the body develops, was a staple of early modern writing. As Thomas Wright proposed:

If we discourse of those Passions which reside in the sensitiue appetite, it euer first intendeth pleasure and delight, because therewith Nature is most con-tented: from which intention followeth, loue, hatred, ire, and such like: this passion beasts most desire, yea children and sensuall persons wholy seeke af-ter, and direct (almost) their whole actions, thereunto: for pleasure is the polestarre of all inordinate passions.[25]

He added that children "lacke the vse of reason, and are guided by an in-ternall imagination, following nothing else but that pleaseth their senses, even after the same maner as bruite beastes doe."[26] William Kempe, like Bryskett, took this capacity for the ascendancy of the beast further into the life of the human when he wrote in 1588 that "youth is forgetfull, not greatly moued with regard of things past, or things to come, but wholy caried away with that which is before their face."[27]

The dangers of this failure to be truly human—this failure to use the ra-tional soul—are clear, and are more troubling than would seem possible from the simple assertion that humans are born human, that they possess that thing which makes them human at birth. This latter assertion would seem to make the status of the human natural, and reiterate the dominion over the world that was given to Adam in Genesis. But this natural status is flawed, as the images of the present-minded infant and the sensual youth re-

---

[24] Bryskett, *Discovrse of Civill Life*, 50, 53–54, 100, and 43.

[25] [Wright], *Passions of the Minde*, 222–23.

[26] Ibid., 12.

[27] W. K. [William Kempe], *The Education of children in learning* (London: Thomas Orwin, 1588), sig. F^r.

veal: humans can fail to *be* human. They have the essence, but they do not always use it, and this failure makes them worse than the beings they have become. In fact, the Puritans John Dod and Robert Cleaver took this failure to its logical extreme in 1630: "it were better for children to be vnborne than vntaught."[28] What this proverb reveals is that birth is not enough, that education plays a crucial role in the rationalizing of the child and the youth. In fact it might be argued that education *makes* the child human.

Despite being born in possession of a rational soul, then, infants are not yet fully human, insofar as human status can only be designated truly by the actions that evidence the possession of a rational soul. An infant exists in a world of merely vegetative desires, without judgment, but with a desire for food and warmth. An infant cannot display reason; it lacks the physiological maturity and capacities (such as movement and, crucially, speech) needed to express it. On this basis, the reasonable nature of infants is a matter of faith, not evidence. As the infant progresses, however, its capacity to begin to express reason does not automatically mean that such a (preexisting) status becomes clear. Rather, things can get worse; the sensitive soul can take over from the vegetative, and desires can become more worldly. What is required to avoid or at least lessen this possibility of the human's becoming merely sensual is a training of the will; is, in short, education.

Creating Humans

According to William Perkins, the first stage in parental care of the infant is baptism,[29] and here godparents make declarations on behalf of the child, declarations that the child, as it matures, is expected to make for him or herself. The reason for the intercession of the godparents is simple: Richard Jones asked in 1583 "Whether maie fooles, mad men, or children bee admitted to the Supper of the Lord?" His answer is clear: "*No, for they cannot examine themselues.*"[30] Self-examination is a key manifestation of human status, and here the infant lacks this capacity but maintains its status as human because the godparents stand in and perform the "self"-examination on the infant's behalf.

Following baptism, parents are expected to continue with their children's godly education, and the catechism—the rote learning of key religious questions and answers—emerges as a primary form of indoctrination.

[28] John Dod and Robert Cleaver, *A Godly Forme of Houshold Gouernment* (London: Thomas Man, 1630), sig. R2ᵛ.
[29] William Perkins, *Of Christian Oeconomie*, in *The Works of that Famous and Worthy Minister of Christ in the Vniuersitie of Cambridge, Mr William Perkins* (London: J. Legatt, 1616–18), 3:694.
[30] Richard Jones, *A Briefe and Necessarie Catechisme* (London: Thomas East, 1583), sig. Cviʳ.

C. John Sommerville has noted that catechismal "memorization . . . was to make orthodoxy the child's second nature."[31] What had been external is made internal; what was culture begins to appear to be nature. In addition, the catechism is a display of key properties of the human—memory and speech. And the latter, according to Dod and Cleaver, could precede the former. "let him haue the words taught him when is able to heare and speake words, and after when he is of more discretion, he will conceiue & remember the sense too."[32] Parroting might appear to dehumanize the child, but in fact the possibility that the child can move from mere repetition to understanding is what makes the distinction: a parrot would always be stuck with parroting. So, in learning the catechism (parents were fined if the child had not learnt it by the age of eight),[33] the child was displaying its entry into the human community as well as into the Christian community. In fact, the two—human and Christian—seem inseparable in this context.

Similar parroting and learning exists in the schools. Here it is not only religious matter that is repeated; Greek and Latin texts are learnt by rote to enable the child to learn rhetoric and classical languages. What is taught to the child is more than merely reading, writing, and so on, however; what is taught at school or at church is, simply put, being human as it is most clearly manifested in the capacity to control one's bodily urges. This process enables children to display their rational natures, to ensure the dominion of will over passion. In his 1634 Reformed household manual William Gouge wrote that "as God hath given to man a reasonable soule, and understanding head, capacity, docility and aptnesse to learne, so ought parents to make use of those parts and gifts, lest for want of using them, in time they be lost: and so children prove little better then bruits."[34] The child's descent into beastliness is a constant threat in treatises, and it is the avoidance of this and the achievement of human status that is the aim of education. The impact of such an education is immense: Dod and Cleaver write, "we are changed and become good, not by birth, but by education."[35] Here is a sense not of development but of metamorphosis, of becoming a new species.

This raises the dangerous possibility that being born with a rational soul is not enough, that there is something more required than the mere possession of this essence of the human. In this regard, two different notions of

[31] C. John Sommerville, *The Discovery of Childhood in Puritan England* (Athens: University of Georgia Press, 1992), 136.

[32] John Dod and Robert Cleaver, *A Treatise of the Exposition Vpon the Ten Commandments* (London: Thomas Man, 1603), sig. 7ʳ.

[33] Sommerville, *Discovery of Childhood*, 145.

[34] William Gouge, *Of Domesticall Duties, Eight Treatises* (London: George Miller for Edward Brewster, 1634), 537.

[35] Dod and Cleaver, *Godly Forme*, sigs. S8ʳ⁻ᵛ.

the human seem to exist. On the one hand, there is the comforting notion of the human as a given status; as a being born and not made. By implication, this status is from God and therefore cannot be undone by humans, whatever actions they perform and however apparently animal-like their behavior. On the other hand, there is the human that is constructed by education, custom, and culture, and this human is, inevitably, reliant on the existence of education, custom, and culture. There is also, of course, a gap between these two conceptions, and with that gap arises the possibility of a (human) being who fails to become a fully fledged (that is, active) human, and whose actions therefore cannot be judged according to the criteria established concerning reason.

The preexistent state of the human, then, is not enough to establish without question the category "human." A belief in the natural difference between human and animal that emerges from the belief in the preexistence of the rational soul does not mean that that difference between human and animal can always be found. And it is the gap between these two categories of the human—the natural and the cultural—that allows for actions which suggest that some humans are not human at all; that some humans are animals. What emerges in the discourses and actions of colonialism, for instance, reproduces some of the arguments surrounding the status of the child.

Colonized Being

If humans have the capacity to give the will dominion over the passions, then animals by implication lack this capacity and are trapped by the body and by immediate, material circumstances. The Italian physician and astrologer Girolamo Cardano offers an image for the enslavement of animals that taps into the common metaphor of the civil war being raged between reason and the passions within each individual, and in so doing gives a vision of the wider implications of this debate. It is worth quoting at length from the English text, *Cardanus Comforte* (1573):

> But this is a thing most proper vnto vs men, that wee shoulde commaunde our selues. For the vertue within vs, moueth our lymmes, because it commeth from vnderstandinge (is ruled wyth a straunge and forraine rule, doth always obaye after one sorte, and is not oure owne simplye, not knowen vnto vs, but we vse it not knowinge, howe we vse it. And so of those thinges whiche come from other where, we be not full maysters of them. So beastes because they be gouerned by the motion of the natural power, and sence, which hath an outward or forraine cause, in like sort be quite voyde of libertye, and vtterly

subiecte to an others gouernment, nothing differing in theyr affections from ŷ sense and seruice which the members in man are wont to do vnto the wil. For if those members be hurt, of theyr owne accorde, & without the commaundement of wyll, they shrinke backe, althoughe they know not wherfor they so doe.[36]

There are three modes of movement: natural, sensitive, and voluntary. According to Robert Burton, natural movement depends "not of Sense," and the example he gives is of a stone falling downward; it is solely of the body. When a limb is injured it moves, Cardano argues, "without the commaundement of wyll": the swift pulling of the hand from the flame is not a reasoned response but a natural one, and is led by the body and not the understanding. Sensitive movement, the second mode, relies on the work of the senses, and is "common to Men and Brutes." Voluntary movement—"a curbe vnto" natural and sensitive movement—is, for Burton, particular to humanity, and not shared with animals. It is a command of will over passion, of mind over matter; it is what Cardano calls the "vertue within vs."[37]

It is the fact that humans have all these possible prompts to movement— natural, sensitive, and voluntary—that distinguishes them from animals. Animals, in Cardano's evocative phrase, are "vtterly subiecte to an others gouernment": they are dependent on the outside world ("gouerned by the motion of the natural power and sence") and forced to respond to it without will or understanding. They are, in this interpretation, colonized by the external, without weapons to mount a defense. Humans too can be colonized, of course, but they can fight that colonization with the weapon of will.

Cardano continues, and underlines the difference between human and animal:

> Moreover and if vnderstandinge were without vs, we shoulde no more differ from other lyuing creatures, then they do one from an other, and nedes it muste folow, ŷ bruite beastes should not want vnderstandinge. Forasmuch as in the same maner the nature both of bruite beastes and men should be illumined in yᵉ same sort, & of the same eternal causes. And nowe it is shewed how brutishe lyuing creatures are for euer, by no kinde of meanes able to attain vnto euē the least shadow of that part which is reasonable, but by memory, or els nature somtime to haue geuen a certain show of some conceiued reason.[38]

---

[36] Girolamo Cardano, *Cardanus Comforte translated into Englishe* (London: Thomas Marshe, 1573), sigs. Cvi^r–v.

[37] Robert Burton, *The Anatomy of Melancholy* (Oxford: Henry Cripps, 1624), subsec. 8.

[38] Cardano, *Cardanus Comforte,* sig. Cvi^v.

The opposition is absolute: animals on one side, humans on the other. And the difference between humans and animals can be found most clearly, Cardano proposes, when considering animals' lack of reason. Animals may *appear* reasonable, he argues, in that they can memorize that which seems reasonable, but memorizing is not the same as *understanding*: "if you trauayle to learne [animals] to vndoe a knot," he writes, "they keepe in memorye how they maye drawe and slake, and so fynallye loose the knot: but if you chaunge the knot neuer so little, they shal neuer know how to vndoe it, vnlesse it be mere chaunce, so as you may well perceyue they be vtterlye deuoyde of reason."[39] The repeated undoing of a single knot is a learned response, not, as the undoing of any knot would be, the action of an understanding mind. This latter capacity would involve discernment, judgment, and abstraction—the capacity to transfer a specific technique to a general problem. Without this capacity to discern, judge, and deal with the abstract, Cardano argues, humans would be the same as animals.

The image of the "forraine rule" of animals by the external world, of animals "subiecte to an others gouernment," provides a link between two aspects of early modern culture that were, in so many ways, deeply and imaginatively intertwined in the period: colonial rule over the "uncivilized" and reasonable rule over the body. But the link between human, animal, and colonialism went beyond this metaphorical representation of the control exercised over the material world. In the early seventeenth century, English writers faced a new problem that called on the debates already outlined in this chapter. If there is a link between colonialism and self-control and knowledge, there is also a link between the colonized and those beings lacking in self-control and knowledge, animals. Cardano's metaphor of rule becomes real when the explorations of the external world expand in their prospect. But while the landscape might change, what remains the same is the debate: What is a human?

Unsafe Possession

The conventional representation of America as a woman plays into the belief that women lack complete (masculine) self-control. When, in "Elegy 19," John Donne writes of his mistress "O my America, my new found land, / My kingdom, safeliest when with one man manned . . ."[40] he is calling on the sexualized language of colonialism itself (and there is a nice irony, as

---

[39] Ibid., sigs. Ciii^{r-v}.

[40] John Donne, "Elegy 19," in *The Complete English Poems,* ed. A. J. Smith (London: Penguin, 1986), 125.

Thomas Laqueur notes, that "in 1559 . . . Columbus—not Christopher but Renaldus—claims to have discovered the clitoris"[41]). The woman, like the uncharted land, requires masculine rule to be fully able to function. The notion of "safeliest" implies the end to the danger of lack of true rule; the end of unruliness.

But the connection of the land with the passionate, the woman, and the body was only one way in which English ideas about the New World and about self-control were brought together. A link was also made between the native peoples of the New World and animals. While the land was (logically) figured as a woman in need of masculine rule, the natives' species status was somewhat more difficult to discern, and this lack of clarity, again logical within the context of ideas about reason, offered some of the English colonists just the excuse they had been looking for.

When the members of the Virginia Company first set out for the New World in 1609 they went supported by the words of numerous sermons and tracts, and the rationale for the colonial expedition was made clear in one anonymous propagandist pamphlet from 1610, *A True and Sincere declaration of the purpose and ends of the Plantation begun in Virginia:*

> The *Principall* and *Maine Ends* . . . weare *first* to preach, & baptize into *Christian Religion*, and by propagation of that *Gospell*, to recouer out of the armes of the Diuell, a number of poore and miserable soules, wrapt vpp vnto death, in almost *inuincible ignorance;* to endeauour the fulfilling, and accomplishment of the number of the elect, which shall be gathered from out all corners of the earth. . . . Secondly, to prouide and build vp for the publike *Honour* and *safety* of our *gratious King* and his *Estates* . . . some small Rampier of our owne.[42]

Historical debate about whether the missionary endeavor really preceded the economic impetus for colonialism still continues, but what is clear from contemporary writings is that the religious ideas were constantly propounded in propaganda.

In the same year as the publication of the anonymous *True and Sincere declaration,* for example, William Crashaw stated the primacy of religious ideas to the Virginia Company once again: "the principal ends there of being the plantation of Church of English Christians there, and consequently the conuersion of the heathen from the diuel to God."[43] Taking the land was

[41] Thomas Laqueur, *Making Sex: Body and Gender from the Greeks to Freud* (Cambridge, Mass.: Harvard University Press, 1990), 64.

[42] *A Trve and Sincere declaration of the purpose and ends of the Plantation begun in Virginia* (London: I. Stepneth, 1610), 2–3.

[43] William Crashaw, *A Sermon Preached in London before the right honourable the Lord*

apparently not the first priority. But, as part of the missionary work, taking control of the terrain was also doing God's work, was returning to him that which had gone astray and had become wild (unmanned). But, obviously, taking that which does not belong to one is theft, and committing such a crime (and sin) would clearly go against the apparently religious underpinnings of this colonial enterprise. How, then, to simultaneously convert the people and the land? How could colonial expansion sit comfortably alongside missionary work? There was an answer, an answer that applied, somewhat horrifically, the logic already outlined in this chapter.

An animal (like most women) does not have the right to own property; animals can only *be* property. Possession signifies control, and control is willed, and therefore available only to a human. A wild rabbit, for example, does not own its burrow but merely lives in it; the land remains in the hands of the human owner. A similar argument was taken up by some propagandists for the Virginia Company. Robert Gray offers perhaps the most succinct version of this argument. Writing in 1609 he states the case for the English mission to the New World clearly:

> yet this earth, which is mans fee-simple by deede of gift frō God, is the greater part of it possessed & wrongfully vsurped by wild beasts, and vnreasonable creatures, or by brutish sauages, which by reason of their godles ignorance, & blasphemous Idolatrie, are worse then those beasts which are of most wilde & sauage nature.

Just as the human will may be swamped by passions and the human fall into a bestial state, so the land too can be overrun by animals and become wild. In a wonderful reversal of the English fear of theft—of illegally taking land belonging to the natives—colonization is figured as God's work. It is a returning to Him of that which has been wrongfully taken; usurpation, of course, is never divinely ordained.

Gray continues, and underlines his point:

> The report goeth, that in *Virginia* the people are sauage and incredibly rude, they worship the diuell, offer their young children in sacrifice vnto him, wander vp and downe like beasts, and in manners and conditions, differ very little from beasts, hauing no Art, nor sciēce, nor trade, to imploy themselues, or giue themselues vnto, yet by nature louing and gentle, and desirous to imbrace a better condition.[44]

LAWARRE, Lord Governour and Captaine Generall of Virginia (London: William Welby, 1610), sig. C3ʳ.

[44] Robert Gray, *A Good Speed to Virginia* (London: Felix Kyngston, 1609), sigs. Bᵛ and C2ᵛ.

Gray's representation of the native people not owning the land on which they live but merely "wandering" on it is not new. Emerging out of Aristotle's definition of the "natural slave," such depictions of New World natives had already been used by European colonists earlier in the sixteenth century. Spanish jurist Diego de Covarrubias wrote in 1547, for example, of Aristotle's words describing "men created by nature to wander aimlessly through the forests, without laws or any form of government, men who are born to serve others as the beasts and wild animals are."[45] Similarly, French cosmographer André Thevet wrote in 1558 that the native peoples of the New World were "a remarkably strange and savage people, without faith, without law, without religion, without any civility whatever, living like irrational beasts, as nature has produced them, eating roots, always naked, men as well as women."[46] For the European colonists these people are like the animals that also roam the land. Ownership, it would seem, goes along with art and science as a truly human trait. However, it is a human trait that the natives lack, and so the natives also lack true humanity. They are beasts, so their land is not private property and cannot, by logical extension, be stolen. The possible double meaning of *human*—its status as a natural and a cultural category—offers the European colonialists a solution.

The fear of the illegality of the colonial enterprise may seem to raise something of a nice theological argument, but the niceties of the argument were felt deeply by those involved in debates in England. Crashaw repeated the fear, and its solution, a year after Gray: he wrote of "not only the lawfulnesse, but euen the excellencie and goodnesse, and indeed the plain necessity (as the case now stands) of this present action."[47] In 1613, three years later, he repeated this theme in his "Epistle Dedicatorie" to a tract by Alexander Whitaker, the minister who traveled over to Virginia with the first voyage. Crashaw wrote that "God himselfe is the founder, the favourer of this Plantation" (thus, perhaps, abdicating English responsibility), adding for good measure that the pilgrims had not traveled over the ocean from want of money but "frō yᵉ extraordinary motiō of Gods spirit." The first

---

Virginia DeJohn Anderson argues that possession and maintenance of livestock provided the English colonialists with evidence of the natives' lack of status. For the English, the fact that the natives had no domesticated animals reflected their lack of civility. Anderson, *Creatures of Empire: How Domestic Animals Transformed Early America* (Oxford: Oxford University Press, 2004).

[45] Diego de Covarrubias cited in Anthony Pagden, *The Fall of Natural Man: The American Indian and the Origins of Comparative Ethnology* (Cambridge: Cambridge University Press, 1982), 96. Pagden's book provides an important overview of Aristotle's ideas as they were taken up in the Renaissance.

[46] André Thevet cited in Olive P. Dickason, *The Myth of the Savage and the Beginnings of French Colonialism in the Americas* (Edmonton: University of Alberta Press, 1997), 30.

[47] Crashaw, *Sermon Preached in London*, sig. C3ʳ.

colonists he termed "the *Apostles of Virginia*."[48] These men were not thieves but evangelists, spreading the word of God to the heathen. For Ralphe Hamor, writing in 1615, the colonists had an even more glorious position. He wrote:

> when these poore Heathens shall be brought to entertaine the honour of the name, and glory of the Gospell of our blessed Sauiour [they shall] . . . cry with rapture . . . Blessed be the King and Prince of England, and blessed be the English Nation, and blessed for euer be the most high God, possessor of Heauen and earth, that sent these English as Angels to bring such glad tidings among vs.[49]

Such arguments would seem to silence any accusations of theft and to present a clear case in support of colonialism. But the simultaneous figuring of the natives as lacking humanity and the missionary zeal of the Company work against each other. If the natives are animals, wandering over, rather than owning, the land—a status that allows the English to take the land without committing theft—how can those same natives be converted? That is, if animals lack the crucial rational, immortal soul, how can they be made into Christians? This paradox sits at the heart of the propaganda of the Virginia Company, and when it is expounded on, it reveals yet another terrifying—and logical—outcome: not only are the natives not human, but neither are the English in any inevitable, timeless, way.

Converting Humans

In her important study *Early Anthropology in the Sixteenth and Seventeenth Centuries,* Margaret T. Hodgen proposes that in this period "peoples are differentiated from one another as 'nations,' while the term 'race' carried a zoological connotation properly applicable only to animals." This taxonomic terminology, she argues, would mean that "As long as man—even pigmented man—was regarded as monogenetic in origin and homogeneous in descent, he could not be submitted to zoological divisions, or to the terms used to designate them."[50] That is, as long as the narrative of creation found in Genesis was adhered to—as it had to be—then there was no possibility of

---

[48] William Crashaw, "Epistle Dedicatorie," to Alexander Whitaker, *Good Newes From Virginia* (London: F. Kyngston, 1613), sigs. A4ᵛ, B4ʳ, and C2ʳ.

[49] Ralphe Hamor, *A True Discourse of the Present Estate of Virginia, and the successe of the affaires there till 18 June 1614* (London: W. Welby, 1615), sig. A4ʳ.

[50] Margaret T. Hodgen, *Early Anthropology in the Sixteenth and Seventeenth Centuries* (Philadelphia: University of Pennsylvania Press, 1971), 214.

proclaiming that the native peoples of the New World were not human. If there were only two original parents, then all humans must come from the same original seed: there must be a relationship between, say, Robert Gray and those natives he depicted wandering the land in Virginia. To claim otherwise would be to propose either that Genesis did not represent the whole story of creation (a blasphemous thought) or that God could make animals that appeared to be human and therefore to undermine the absolute dominion that humanity so regularly claimed from the scriptures. If the latter possibility were taken seriously it would mean that the question, who could tell who was a true human? would be even more difficult to answer. But ultimately it seemed as if the natural, original difference of human and animal could not be upheld. This is something that is revealed in some of those English writings themselves. If a substantial difference is to be asserted between English and New World peoples even as the narrative of creation presented in Genesis is adhered to, then monogenesis—shared parentage— must exist alongside something else, something learned.

Crashaw offers an interpretation that would allow for the simultaneous animal status of the natives and their convertibility; it takes us back to the debate about the nature of the infant. He writes:

> for the time was when wee were sauage and vnciuill, and worshipped the diuell, as now they do, then God sent some to make vs ciuill, others to make vs Christians. If such had not been sent vs we had yet continued wild and vnciuill, and worshippers of the diuell.[51]

"If such had not been sent vs": here the English are presented as passive; as the recipients, not the agents, of change; as suffering rather than working.

In similar terms, Alexander Whitaker also emphasized the shared parentage of all humans, and the transformation of the English over time:

> One God created vs, [the natives] haue reasonable soules and intellectual faculties as well as wee; we all haue *Adam* for our common parent; yea, by nature the condition of vs both is all one, the seruants of sinne and slaues of the diuell. Oh remember (I beseech you) what was the state of *England* before the Gospell was preached in our Countrey?[52]

In this discourse the natural, and inevitable, distinction of human from animal, and of human from savage (with the two made one by the common conflation of savages with animals), is once again gone. Truly human status is achieved only over the space of time, only through learning.

---

[51] Crashaw, *Sermon Preached in London*, sig. C4ᵛ.
[52] Whitaker, *Good Newes From Virginia*, 24.

This shift in focus, from Aristotle's idea of the absolute distinction between the natural slave and the human to the recognition of the constructed nature of the human, echoes, once again, a shift that had taken place in the ideas of earlier European colonialists. As Anthony Pagden has noted, to construct an argument for the duty of care that the educated (and Christian) Spanish had over the New World natives, Spanish thinkers "had to generate their own theory of the relativity of human social behaviour." They had, in fact, to shift from "one branch of Aristotelian psychology (concerned with the mental status of slaves) to another (concerned with the mental disposition of children)." The conclusion, Pagden argues, is "a theory that all human cultural behaviour, and nearly all beliefs, are the outcome of social conditioning."[53] A human, once again, emerges as a being not simply born, but made.

But, of course, declaring the natives to be as human as the English colonialists rather undoes the (useful) logic of possession. One outcome of the conversion narrative would be that ultimately, and inevitably, the land should be returned to the natives once they have achieved human status. Just as a ward in the law came into possession only at his or her age of majority, should not the same reasoning be applied to the natives of the New World? Clearly, this thought does not enter the minds of the propagandists of the Virginia Company. And, in fact, what operates for them is not the logic of simultaneous being and becoming human that operates in the English law in relation to wardships. Instead, the English colonialists appear to maintain both narratives—of the natives' lack of humanity, and of their innately human status—while at the same time ultimately regarding the natives, like women, as lesser beings.

In this sense, the notion of the human becomes even more complex. There are natural born humans who can only be human because they possess the rational soul. Then there are humans in possession of the rational soul who require education to become truly human. Finally, there are humans who possess rational souls, can be educated, but are still less human than the human. Thus the category begins to collapse into absurdity. The possession of a rational soul no longer seems to offer a way of distinguishing human from animal. But this might be, perhaps, to misunderstand some of the ways in which the human was represented in the discourse of reason. It might be to fail to recognize that human itself is a vast category that includes not only those beings who exhibit the properties of humanity, those who are born human, those who become human; it also includes those who cease to be human, and it is the capacity to lose one's status—to become a beast—that is the focus of the next chapter.

---

[53] Pagden, *Fall of Natural Man*, 3.

CHAPTER 3

Becoming Animal

In "To Sir Edward Herbert, at Juliers," written in 1610, John Donne offers a helpfully explicit vision of the human relationship to the animal in the discourse of reason, and in doing so reveals the danger of that relationship. Donne writes:

> Man is a lump, where all beasts kneaded be,
>   Wisdom makes him an ark where all agree;
> The fool, in whom these beasts do live at jar,
>   Is sport to others, and a theatre,
> Nor 'scapes he so, but is himself their prey;
>   All which was man in him, is eat away,
> And now his beasts on one another feed,
>   Yet couple in anger, and new monsters breed;
> How happy is he, which hath due place assigned
>   To his beasts, and disafforested his mind![1]

The "disafforestation" of the human mind is the ideal in the achievement of human status. Here, as in the conquest of the New World, the aim is that the wilderness be stripped away and animals put in their proper place. What remains is land fit for use, is reason unencompassed by what Robert Burton termed "ferall passions."[2]

---

[1] John Donne, "To Sir Edward Herbert, at Juliers," lines 1–10, in *John Donne: The Complete English Poems,* ed. A. J. Smith (London: Penguin, 1986), 218. Smith glosses the "theatre" here as a "wild-beast show" (539).

[2] Robert Burton, *The Anatomy of Melancholy* (Oxford: Henry Cripps, 1624), 175.

But in Donne's image of the human as the culmination of all animals—
a logical extension of the idea of man as microcosm—the animal holds a
complex status. It exists as the other against which the human is constructed,
as the figure of lack that reveals the plenitude of the human. Alongside this
sense of the human, however, there exists another in which the human can
cease to be human, can fail to assign due place "To his beasts." If a human
is human because of the expression of reason—the display of the possession
of the rational soul—there exists the prospect that what is expressed by a
human may not after all be reasonable. So, as well as the possibility that human as a category is constructed and not natural—that humans must be created and are not simply born—there exists the risk that humans may cease
to be human, may stop acting according to their education, and may revert
back to their natural sensuality. The logical description of this absence of evidence of reason is a descent to the animal, is the revelation of the "beast in
man."

This threat to human status pervades writings of the period, where animals are simultaneously other and self: they are held up as the things that
humans are not, and yet humans are frequently depicted as becoming them.
In this sense the othering of animals actually breaks down because of the
logic that underlies it. If an animal is the thing that a human is not, and yet
a human can cease to be (or never become) the thing it is, then an animal
is something much more than other: it becomes kin. In early modern England, writers offered a way around this conundrum, but the solution did
not clarify human status, it clouded it. Already confusingly both born and
made, natural and cultural, now humans emerge not as superior to animals,
not even as animals, but as beings who are simultaneously human and animal. What also emerges is a new sense in which humans may not be superior but may actually be inferior to their animal others; and all this by
following the logic of the discourse of reason.

Natural Descent

The descent to the status of an animal is an utterly logical one in the discourse that has been our focus so far. For Thomas Wright the passions "blind
reason, they seduce the wil," and such seduction reveals itself in a kind of
transformation: he argues that passions "changed them like *Circes* potions,
from men into beasts."[3] Likewise, Joseph Henshaw wrote, "Deliberately to
move to any businesse is proper to man; headily to be carried by desire, is

---

[3] Th. W. [Thomas Wright], *The Passions of the Minde* (London: V. S. for W. B., 1601), 3, 100.

common to beasts."[4] And of course to be "headily carried by desire" is something that all humans are capable of. Such representations give a moral status to the descent, in that they make it evidence of a willed but vicious choice on the part of the individual, and as Leonard Lessius wrote concerning intemperance in 1636,

> a disordered life repayes that small and fading pleasure, which it affords to the throat, with an innumerable companie of mischiefs: For it oppresseth the belly with the weight thereof, it destroyes health, it makes the bodie to become noysome, ill-sented, filthie, and full fraught with muck and excrement; it enflames Lust, and enthralls the minde to passions; it dulls the Senses, weakens the Memorie, obscures the Wit and Understanding, &, in summe, makes the Minde become lumpish.[5]

This is a truly vicious circle: lack of temperance reveals itself in overindulgence, from which overindulgence illness ensues. As Michael C. Schoenfeldt has written, in the early modern period "health assumed the role of a moral imperative. . . . Illness in turn was perceived as a symptom of immorality."[6] As such, illness marks the human's lack of virtue, and therefore lack of human status, which, in turn, is revealed in a lack of temperance. In fact, just as Plato, you remember, never laughed and therefore revealed his ultrahuman status, so in 1584 Thomas Cogan told a story that revealed Socrates' almost divine renunciation of the body. Socrates, Cogan wrote, "of set purpose oftentimes exercised and enured himselfe to endure hunger and thirst: which be more hard to suffer than to feede moderately, and to forbeare that which reason forbiddeth, although our appetite desire it. And when he was demaunded why he did so, that I may not accustome my selfe (quod he) to folowe my sensuall appetites, lustes, & desires."[7] Such absolute control is impossible for most humans to follow for a good reason. Thomas Adams cited Martin Luther's explanation, revealing the very real power of the body in a way that reiterates the conventional analogy of self-control and civil order in a particular religious context: "All of vs, as *Luther* was wont to say, haue naturally a Pope bred in our bellies."[8] For the Pope to take charge would be, in

[4] [Joseph Henshaw], *Meditations miscellaneous, Holy and Humane* (London: R. B., 1637), 71–2.

[5] Leonard Lessius, *Hygiasticon: or, the right course of preserving Life and Health unto extream old Age*, 3rd ed. (Cambridge: Printers to the University, 1636), 203.

[6] Michael C. Schoenfeldt, *Bodies and Selves in Early Modern England: Physiology and Inwardness in Spenser, Shakespeare, Herbert, and Milton* (Cambridge: Cambridge University Press, 1999), 7.

[7] Thomas Cogan, *the Haven of Health* (London: Henrie Midleton, 1584), sig. ¶3ᵛ.

[8] Thomas Adams, *The Praise of Fertilitie*, in *The Workes of Tho: Adams* (London: Iohn Grismand, 1629), 1034.

this framework, for the individual to give up self-control in favor of dominion by a deceitful tyrant. Once again, human status is revealed for all but the most extraordinary humans as a struggle.

This framework, wherein temperance is a virtue not merely of diet but of selfhood, is carried over into discussions of other vices in which the danger of descent to the status of an animal is more obvious. What also becomes apparent, however, is that the human, in becoming a beast, remains—apparently paradoxically—absolutely human. When one analyzes the concept of the "beast in man" more closely, in fact, what can be found are not beasts at all but humans.

The Disappearance of the Animal

For George Gascoigne drunkenness is a "beastlie vyce." He wrote in his *A Delicate Diet, for daintie mouthde Droonkardes* (1576): "let mee set downe this for my generall proposition, *That all Droonkardes are Beastes.*" Quoting from an *Admonition* by Saint Augustine, Gascoigne wrote, "in theyr droonkennesse they know neither themselves, nor any body else: neither can they goe, stande, nor speake any thing that pertayneth unto reason."[9] So the status of human as animal seems logical, and it is no surprise to find the following image in Thomas Young's 1617 assault on drunkenness:

> a Forrest is a safe harbour and abiding place of Deere or Beasts, not of any sort whatsoeuer: but of wilde and such as delight in Woods, (and herevpon a Forrest hath the name (as one would say *Feresta*, that is a Station of wilde Beasts: and likewise I thinke the inhabitants of [ale houses], learne their sauage manners, and brutish behauiour, because they conuerse chiefely with Beasts: For they haue no Magistrates, nor they will hire no Ministers, for they goe ten times to an Ale-house, before they goe once to a Church.[10]

Donne called for a "disafforestation" of the mind, and here Young presents the ale house as a kind of woodland, and the inhabitants of the ale house as, by analogy, wild animals. Getting drunk, it would seem, is a failure to assign due place to the beasts of one's mind.

But the transformation of human to animal is not only by analogy. The figure of Circe, who transformed Ulysses' followers into animals, is an im-

---

[9] George Gascoigne, *A Delicate Diet, for daintie mouthde Droonkardes* (1576; repr., London, 1789), iv, 6, and 8.
[10] Thomas Young, *Englands Bane: Or, The Description of Drunkennesse* (London: William Jones, 1617), sig. F^r.

age, for Thomas Wright, of the transformative power of the passions, and
she emerges once again in diatribes against drunkenness to embody the
transformative power of drink.[11] Gascoigne, Thomas Heywood, and R. Junius all refer to Circe in a way that conveys the dangerously demonic quality of alcohol and the fragility of the status of the human in the face of such
a powerful potion.[12] The repetition of the Circe story is absolutely appropriate because in early modern thought such metamorphoses as those undergone by Ulysses' men were understood, in Sir Walter Raleigh's terms, "to
shew the change of mens conditions, from Reason to Brutalitie, from Vertue
to Vice, from Meeknesse to Crueltie, and from Iustice to Oppression. For by
the liuely Image of other creatures did those *Ancients* represent the variable
passions, and affections of mortall men."[13] Thomas Walkington, writing in
1607, likewise presented another such metamorphosis in terms that will be
recognizable by now: Io, transformed into a cow, is described as being "*frõ
her selfe estranged.*"[14] And this sense of self-estrangement is also used in *The
Contention Betweene three Brethren* a year later, in which one of the brothers
proposes that the man given over to drunkenness will "alienate himselfe
from himselfe, that is to say, from his vnderstãding and power, and . . . become without iudgement and reason, like vnto a childe of a yeare old."[15]

The alienation of the self from the self is also given another meaning in
early modern representations of drunkenness in the popular depiction of
different kinds of drunkards as particular sorts of animals. Thomas Young
writes:

> The first is Lyon drunke, which breakes glasse windowes, cals his Hostesse
> Whoore, strikes, fights or quarrels, with either Brother, Friend or Father. The
> second is Ape-drunke, who dances, capers, and leapes about the house, sings
> and reioyces, and is wholly rauisht into iests, mirth and melodie. The third
> is sheepe drunke, who is very kinde and liberall, and sayes, by God captaine
> I loue you? . . . The fourth is Sow drunke, who vomits, spewes, and wallowes
> in the mire, like a Swine. . . . The fift is Foxe drunke, who being of a dull
> spirit: will make no bargaine till hee hath sharpened his wit with the essence

[11] For Circe, see Homer's *Odyssey*, in which Ulysses' followers are presented "Grovelling
like swine on earth, in foulest sort." *The Odyssey*, trans. George Chapman (1614–15; repr.,
Ware: Wordsworth, 2002), Book 10, line 329.

[12] Gascoigne, *Delicate Diet*, 12; Thomas Heywood, *Philocothonista, or The Drvnkard, Opened,
Dissected, and Anatomized* (London: Robert Raworth, 1635), 2; and R. Junius, *The Drunkards
Character, Or, A True Drunkard* (London: R. Badger, 1638), 12–13.

[13] Sir Walter Raleigh, *The Historie of the World. In Five Bookes* (London: Walter Burre,
1614), 27.

[14] T. W. [Thomas Walkington], *The Optick Glasse of Hvmors* (London: John Windet, 1607),
sig. 11r.

[15] *The Contention Betweene three Brethren* (London: Robert Raworth, 1608), sig. D2v.

of good liquor, and is then so craftie and politique, that hee deceiues any man that shall deale with him. . . . The sixt is Maudlin drunke, who weepes, cryes, and whines, to see the goose goe barefoote. The seuenth is Goate drunke, who is in his drinke so lecherous, that hee makes no difference of either time, or place, age or youth. . . . The eight is Martin drunke, which will be drunke betimes in the morning, or alwayes the first in the company, yet will he neuer cease drinking, till he hath made himselfe freshe againe. The ninth is Bat drunke, which are a sort of *Drunkards* that will not openly be seen in such actions, but as the reremowse or Bat, delights in secret places and flies by night.[16]

Thomas Heywood, writing in 1635, repeated this sense of drunkenness as an animalization of the human,[17] and figured, on the title page of *Philocothonista,* an image of such animalized drunkards (see figure 3.1).

It is interesting to note in the light of discussions of gender in the previous chapter that in this image the female servant has maintained her human shape (and thus, in the context of this representation, status) while the male drinkers have become animals. The image thus depicts degeneration taking place on an individual basis as well the destruction of hierarchy on a social one. By giving precedence to the body, by overindulging in alcohol, and by destroying the immaterial capacity—Gascoigne writes the drunkenness will "weaken reason and understanding"[18]—the drinker logically becomes an animal, and by consequence the social order into which the human should fit also breaks down.

But this is not the end of the story of the drunkard as beast. Being drunk is not simply "voluntary *madnesse*," as Thomas Adams termed it; individuals do not merely and willfully remove the qualities that make them human, and therefore effect a transformation to the status of the beast. It is worse than that. Adams says of the drunkard, "Nay, he is in some respect worse then a beast: for few beasts will drinke more then they need, whereas *madde Drunkards* drinke when they haue no need, till they haue need again."[19] Likewise, Gascoigne noted that "Beasts doo by a natural enstincte observe a certaine mediocritie, in many thinges whiche doo by extremitie turne into vice." And Junius noted in 1638 "there is little difference between [a drunkard] and a beast, but that he doth *exceede a beast* in *beastlinesse.*"[20]

<hr/>

[16] Young, *Englands Bane,* sig. F2v–F3r. I am grateful to Adam Smyth for first introducing this animalization of the human drunkard to me.

[17] Heywood, *Philocothonista,* 2–6.

[18] Gascoigne, *Delicate Diet,* 20.

[19] Thomas Adams, *Mystical Bedlam, or The World of Mad-Men* (London: George Purslowe, 1615), 61.

[20] Gascoigne, *Delicate Diet,* 15; Junius, *Drunkards Character,* 2–3.

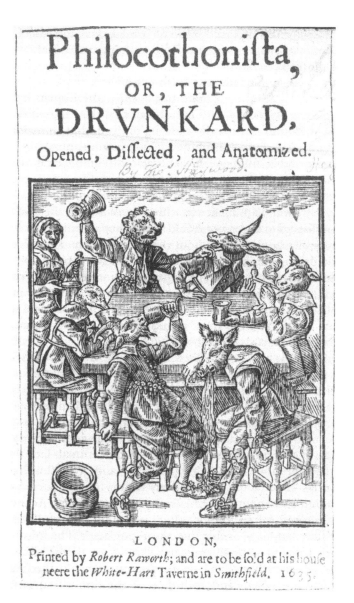

Figure 3.1. Title page, from Thomas Heywood, *Philocothonista, Or, The Drunkard, Opened, Dissected, and Anatomized* (London: Robert Raworth, 1635). By permission of the British Library, classmark C30d11.

framework, as we have seen, animals are initially represented as the absolute other. Despite the prospect of the human becoming a beast, animals are perceived to have no community with humans; they are the things *against* which humans position themselves. But as well as severing the ties between humans and animals, inward government theory also works to undercut the notion of human community. In discussions of cruelty, for example, writers are not interested in the suffering of the moral patient—the individual enduring the cruelty. Instead, writers focus on the moral agent—the individual being cruel—and thus self-control, not suffering, is central. This is something that can be clearly traced in a text that had a massive influence on Renaissance thinking about cruelty: Seneca's *De Clementia*.

Seneca's work, like that of so many classical thinkers, was widely available in England, and was translated into English by Thomas Lodge in 1614 as *A Discourse of Clemencie*. In it Seneca states, "Crueltie is humane euill, it is vnworthy so milde a minde: this is a beast-like rage to reioice in bloud and wounds, and laying by the habite of a man, to translate himselfe to a wilde beast."[23] Seneca's representation of the effect of being cruel makes no mention of the individual suffering the infliction of cruelty because the focus is on the active individual's struggle to maintain reason.

The nature of the effect of cruelty that Seneca describes should be recognizable by now. Cruelty, a vice, is a property of the human, but in being so it reveals the frailty of human reason; cruelty makes a human like a beast in that it transforms the human into something other than the rational self. But, of course, cruelty is a "humane euill"; it is not something that an animal is capable of, and therefore the wild beast that the cruel human is transformed into is not an animal at all; it is, in fact, a human.

This perception of cruelty found its echo in numerous early modern writings. The early sixteenth-century humanist Juan Luis Vives, for example, proposed that "it is inhuman and cruel when we relinquish any human judgement and feeling to adopt that of the beasts."[24] And, in a similar vein, in 1625 Robert Wilkinson argued that "some men endeauour by crueltie and couetousnesse, not onely to kill humanitie and kindnesse in their hearts, but euen to depopulate, and roote out mankinde it selfe from the earth."[25] By being cruel, humans destroy other humans, but more significantly in this discourse, they destroy their own humanity and descend to the status of the beast. Paradoxically, this descent is something only humans can

---

[23] Seneca, *A Discourse of Clemencie*, in *The Workes of Lucius Annæus Seneca, Both Morall and Naturall*, trans. Thomas Lodge (London: William Stansby, 1614), 601.

[24] Juan Luis Vives, *The Passions of the Soul*, trans. Carlos G. Noreña (Lampeter: Edwin Mellen, 1990), 92.

[25] Robert Wilkinson, *The Stripping of Joseph, Or the crueltie of Brethren to a Brother* (London: Henry Holland, 1625), 3–4.

achieve. Animals, it would seem, are wild but not cruel; or, if they are de-picted as being cruel then that cruelty is in this logic an anthropomorphic, not (to use an anachronistic term) a zoological, statement. Hence William Gouge is able to write what seems utterly paradoxical: "the most cruell beasts that be are very tender. . . ."[26] There is a difference between being an ani-mal and being beastly. Beastliness is not a metaphorical representation of a human action; it is a true description, in that being cruel is an abdication of reason, a failure to use the faculties of the rational soul. The beast in beast-liness is always human, you might say; is always other than an animal.

However, cruelty is not only found in action—in being cruel—it can be traced in inaction as well. Charles Fitz-Geffry's *Compassion Towards Captives* (1637) deals with the very real situation of Englishmen imprisoned in "Bar-barie." His focus is not on the cruelty of the Turks but on the lack of com-passion of the English living at liberty in England; in fact, for him those two positions are not dissimilar. Lacking compassion, the free English are taking no action to help stop the cruel treatment of their compatriots, and thus they are themselves being cruel. The focus of the text appears to go beyond the self of inward government theory by looking at the imprisoned other, but in reality it is the self—the free Englishman—who is the focus.

Beginning with Hebrews 13:3—"Remember those that are in bonds as bound with them"—Fitz-Geffry writes of the prisoners: "Their *passion,* our *compassion*. Their misery is bondage and captivity: They are in bonds: Our duty is to extend unto them a twofold mercy; I. *Consideration;* we must *re-member them;* 2. *Compassion;* we must so remember them as if we our selves were *bound with them*." At one point he presents the human body—in which the tongue will speak out, giving voice to pain, even though the pain is felt in the toe—as the natural (and therefore truthful) exemplar of compassion, and reiterates the notion of synderesis by placing the innate presence of a notion of right and wrong in the human.[27] Again, versions of this can be found elsewhere. Writing of Revelation 1:9, the Elizabethan Calvinist William Perkins states:

> [John] calls himselfe their *copartner in affliction;* because his pitifull heart was mooued with the bowels of compassion, towards all his fellow members; when hee remembred their persecution and affliction vnder the cruell tyrant *Domitian*. And the same affection should bee in euery one of vs towards the poore afflicted seruants of Christ: seeing they be our fellow members, we should haue a fellow-feeling with them.

[26] William Gouge, *Of Domesticall Duties, Eight Treatises* (London: George Miller, 1634), 537.
[27] Charles Fitz-Geffry, *Compassion Towards Captives* (Oxford: Leonard Lichfield, 1637), 3, 24.

As Fitz-Geffry does in his later text, Perkins uses the analogy of the body to emphasize the naturalness of compassion.[28]

So, lacking compassion, whatever its form—active or passive—is cruel, a lack of human feeling; and, as Fitz-Geffry writes of the English who feel no compassion for their imprisoned compatriots, "From our selves desend we to bruite beasts."[29] One makes oneself beastly by being cruel, but one also makes oneself beastly by doing nothing. Unreasonable thoughts and no thoughts at all have equivalent destructive power; thoughtlessness is, after all, a lack of thought.

But this tale of human cruelty takes yet another turn. Not only does the failure to maintain a fixed and reasonable status facilitate being cruel (for if reason were always at work, cruelty would not exist); it also facilitates being treated cruelly. The moment of cruelty does not only reveal the agent's vicious status; it also degrades the victim, the moral patient. So great is the concentration on the self that the selfhood of others—even human others—disappears. In 1609 Robert Tylney recorded Tacitus's tale of "The Saints of God suffering martyrdome" in which the martyrs were "made laughing stockes to the world; in being clad with skinnes of wilde beasts, and so exposed to the tearing and renting asunder by dogges."[30] Here cruelty is represented as an artful addition; the martyrs are not merely punished for their beliefs but are also subjected to a kind of "sport" that is, to say the least, cruel and unusual (a phrase that entered the English law in 1689).[31] What is also important in Tylney's representation of this martyrdom is that, of course, the martyrs are made animal or, rather, animal-like (they are costumed), and this making-animal of the victim of cruelty recurs again and again in the early modern period. In fact, it appears to allow for cruelty to be inflicted: in Tylney's words, the martyrs were dressed in skins "and so" thrown to the dogs; there is cause and there is effect.

But this making-animal of the victim is, like the descent of the agent of cruelty to the status of an animal, merely figurative. Just as the agent, even when acting like a beast, remains firmly human, so the patient is only symbolically made animal or, to use Tylney's image, disguised as an animal. However, it is this disguise (and, of course, it need not be as literal as it is in Tylney's tale) that allows for cruel treatment. A canonical version of this double descent can be traced in act 3 of Shakespeare's *King Lear* (c. 1605), at the blinding of Gloucester. As the scene opens and Gloucester is yet to appear on the stage he is firmly human—he is described by Cornwall as a trai-

---

[28] William Perkins, *An Exposition upon the I Chap. Of the Revelations*, in *Works*, 3:235.

[29] Fitz-Geffry, *Compassion Towards Captives*, 24, 22.

[30] Robert Tylney, *Two Learned Sermons* (London: W. Hall, 1609), 9.

[31] See J. H. Baker, *An Introduction to English Legal History*, 3rd ed. (London: Butterworths, 1990), 584.

tor, a thief (only humans can be treacherous or can possess or steal private property). However, as soon as Gloucester enters the stage his status shifts: he is animalized and thus can be victimized. Regan says, "Ingrateful fox, 'tis he." Twenty-seven lines later, Gloucester himself reiterates his new status: "I am tied to th' stake, and I must stand the course." His capacity to speak his descent reminds the audience of the absolute cruelty they are witnessing: he is a human in the position of an animal. In addition to this, and as if to reiterate this point, Gloucester also sets in motion the second descent found in discussions of cruelty: just as he is being animalized in order to be brutally tortured, so the torturers also become animal. When asked by Regan why he has sent the exiled King to Dover, Gloucester answers,

> Because I would not see thy cruel nails
> Pluck out his poor old eyes, nor thy fierce sister
> In his anointed flesh stick boarish fangs.

The sisters become animals in their ferocious cruelty to their father, and Regan's valediction to the blinded Gloucester merely reinforces his status and her lack of compassion: "Go thrust him out at gates, and let him smell / His way to Dover."[32] Gloucester has become a sensing, not a reasoning, being.

The figurative making-animal of moral patients is something that comes up again thirty two years later in Fitz-Geffry's discussion of the imprisonment of Englishmen by Turks. The Turks, he writes, treat the Englishmen with outrageous and animalizing cruelty: their markets are, apparently, "*fuller of our men then ours are of horses and cattle: Christians there bought, sold, cauterised, seared, as wee doe beasts.*"[33] But just as cruel humans remain human in their beastliness, so the figurative nature of the descent of the victim to the status of an animal is made explicit when writers note that human victims actually have less status than animals. As Cordelia says of the torments that Lear has undergone at the hands of his "pelican daughters": "Mine enemy's dog, though he had bit me, should have stood / That night against my fire."[34] Similarly, Fitz-Geffry writes that the imprisoned English "are men, should we see a man beating his horse, his dog, as our men are beaten by these *circumcised dogs,* wee would pity the poore beast and crie out that the owner were a verier beast then that he beateth."[35] The fact that no such cries have been heard, he implies, shows that both the Turks and the English at home regard the imprisoned Englishmen as having less status than animals.

---

[32] Shakespeare, *King Lear,* 3.7.22–23, 3.7.26, 3.7.52, 3.7.54–56, and 3.7.91–92.

[33] Fitz-Geffry, *Compassion Towards Captives,* Preface, n.p.

[34] Shakespeare, *King Lear,* 3.4.71 and 4.6.29–30. "Pelican" daughters here refers to the belief that the pelicans fed their children with their own blood.

[35] Fitz-Geffry, *Compassion Towards Captives,* 43.

However, just as real (as opposed to symbolic) animals disappear in discussions of cruelty, so the animals that are used to mark the absolute decimation of the status of the victim of cruelty remain, I think, symbolic. When Cordelia claims she would have let a dog sleep by the fire on such a stormy night, it is not evidence of kindness to animals so much as an emphatic disavowal of the absolute cruelty of her sisters' actions that is being voiced. And when Fitz-Geffry refers to the man beating his horse, this, again, is not a real issue in his ethical stance but merely a means of enlarging on the thoughtlessness of the English in regard to other English *men*. This absence of animals from discussions of cruelty is not wholly surprising: according to Keith Thomas "In the case of animals what was normally displayed in the early modern period was the cruelty of indifference. For most persons beasts were outside the terms of moral reference. . . . It was a world in which much of what would later be regarded as 'cruelty' had not yet been defined as such."[36] What Thomas fails to note is that the "cruelty of indifference" was directed at other humans as well. Such indifference was a logical outcome of the nature of the discourse in which the focus was on the individual being cruel rather than on the victim of cruelty, in which the status of the self took priority over the status of the other.

But it is not only in relation to the descent of the human that animals become symbolic. It is also in ethical discussions that seem more clearly and obviously concerned with relationships with real animals that animals are merely symbolic. To be kind to an animal, it turns out, is not to be kind to an animal at all.

Saving the Beast's Life

As well as classical writings by Aristotle, Plato, Seneca, and so on, early modern thinkers were indebted in many ways to the work of the thirteenth-century theologian Thomas Aquinas. Aquinas was instrumental in bringing classical thought together with Christian, and he took from Plato and Aristotle a belief that within God's creation there is a chain of being which organizes that world. Arthur O. Lovejoy, quoting from Aristotle, has defined such a "conception of the universe" as one in which there was

> an immense, or . . . an infinite, number of links ranging in hierarchical order from the meagerest kind of existents, which barely escape non-existence, through "every possible" grade up to the *ens perfectissimum*—or, in a somewhat more orthodox version, to the highest possible kind of creature, be-

[36] Keith Thomas, *Man and the Natural World: Changing Attitudes in England, 1500–1800* (London: Penguin, 1984), 148.

tween which and the Absolute Being the disparity was assumed to be infinite—every one of them differing from that immediately above and that immediately below it by the "least possible" degree of difference.[37]

Human superiority to animal is, of course, based on possession of reason, while animal superiority to plant is based on the capacities for movement and perception (capacities that come from the possession of the sensitive soul). Both of these forms of superiority are presented as natural and are evidenced in use: Aquinas states "It is, therefore, legitimate for animals to kill plants and for men to kill animals for their respective benefit." However, while in the chain of being there is a categorical difference between plants and animals, in the human moral relationship to the natural world this distinction disappears. Aquinas writes: "He who kills another's ox does indeed commit a sin, only it is not the killing of the ox but the inflicting of proprietary loss on another that is the sin. Such an action is, therefore, included not under the sin of homicide but under that of theft or robbery." Killing an ox is little different from, say, stealing a cart: butchering a farmer's animals or destroying his crops would have equal status before God.

However, even as he appears to present them as mere objects, there is in Aquinas's theory the possibility of kindness to animals, but this kindness does not represent a vision in which animals are conceptualized as fellow beings in the world—far from it. Animals, Aquinas writes, can be "loved from charity as good things we wish others to have, in that by charity we cherish this for God's honour and man's service."[38] That is, animals should be cared for not for their own sake but for the sake of their owners, or of God. This perception of animals is taken up in England in the early modern period.

Taking their lead from Aquinas and from Proverbs 12:10—"a righteous man regardeth the life of his beast"—numerous early modern theologians were led into discussions of the moral responsibility of humans toward animals, but their discussions remained—logically—uninterested in animals. Writing in 1589, Thomas Wilcox stated: "hee is mercifull, if to beastes, much more to men."[39] Likewise, in 1592 Peter Muffett wrote, "if he be so pitifull to his beast, much more is he mercifull to his seruants, his children, and his wife."[40] Here we have a glimpse of the natural world in microcosm, of a do-

---

[37] Arthur O. Lovejoy, *The Great Chain of Being: A Study of the History of an Idea* (1936; repr., Cambridge, Mass.: Harvard University Press, 1976), 59.

[38] Thomas Aquinas, *Summa Theologiae*, trans. R. J. Batten (London: Blackfriars, 1975), 38:21, 34:,91

[39] T. W. [Thomas Wilcox], *A Short, Yet sound Commentarie; written on that woorthie worke called; The Prouerbes of Salomon* (London: Thomas Orwin, 1589), sig. 38ʳ.

[40] P. M. [Peter Muffett], *A Commentarie Vpon the Booke of the Prouerbes of Salamon* (London: Robert Field, 1592), 103.

mestic chain of being: animals are at the bottom, with the master–father–husband at the top. Even in this inferior position, though, animals are still perceived to be within the moral compass of humanity, but for a particular reason: becoming inured to viciousness to animals, so the Thomist argument goes, makes one more likely to be vicious to humans, something that would endanger not only other humans (a concern, but not the most important one here) but also one's own immortal, rational soul (the greatest concern of all). In his 1612 sermon *Mercy to a Beast* John Rawlinson put this view succinctly: "Saue a beasts life, and saue a mans."[41]

Keith Thomas has labeled this early modern perception of animals as beings within the moral compass of humanity as a "new attitude," and argues that it is paradoxical that such a vision should come from "the old anthropocentric tradition."[42] What he fails to take full notice of is that these "new" ideas not only merely repeat what can be found in the much older Thomist model but also remain absolute in their concentration on the human self. Kindness to animals is given consideration not because animals deserve to be treated with kindness but because such kindness is self-serving. As Joseph Hall wrote, "The mercifull man rewardeth his owne soule; for Hee that followeth righteousnesse and mercy, shall find righteousnesse, and life, and glory; *and therefore,* is blessed *for ever.*"[43] It is the reward of the self that is the focus of apparently ethical actions toward animals, and one logical outcome of this is that, once again, animals are merely symbols (albeit living and breathing ones) through which the self can contemplate its inner being. To think about animals as subjects of compassion, as beings worthy of kindness in and of themselves, would undo the discourse of reason. This is something that is staged in Shakespeare's *As You Like It* (c. 1599). Whereas in *King Lear* Shakespeare reiterates the notion of cruelty that was central to inward government theory, in this earlier play he sets up an alternative vision of ethics only to crush it.

Feeling Transformations

In *As You Like It* Jaques' "weeping and commenting / Upon the sobbing deer" leads him to moralize about the status of those humans who perform the killing: "he swears you do more usurp / Than doth your brother that hath banished you," says the First Lord to Duke Senior, whose hunters have hurt the deer. The tyranny of Duke Frederick is as nothing, in Jaques' mind, compared to that shown by the hunting party:

---

[41] John Rawlinson, *Mercy to a Beast* (Oxford: Joseph Barnes, 1612), 33.
[42] Thomas, *Man and the Natural World,* 156.
[43] Hall, *Salomons Diuine Arts,* 64.

Thus most invectively he pierceth through
The body of the country, city, court,
Yea, and of this our life, swearing that we
Are mere usurpers, tyrants, and what's worse,
To fright the animals and to kill them up
In their assigned and native dwelling place.[44]

Human cruelty—hunting—is figured as an unnatural assault, as a failure of "imaginative dwelling" in the native dwelling of animals.[45]

However, Jaques' compassion is represented by the First Lord as a failing, and in this the First Lord becomes the voice of reason—the voice of the theory of inward government. When Jaques cannot be found in the forest, the First Lord offers a solution: "I think he be transformed into a beast, / For I can nowhere find him like a man."[46] Jaques' compassion for the stag is figured as transformative; in the eyes of those following the logic of the discourse of reason, he becomes a beast in his fellow feeling with the animal. This is not an actual transformation (Ovid has no place here); rather, the transformation is once again a logical outcome of the ethics of the discourse of reason. In identifying with the animal, Jaques has failed to use the thing that distinguishes him from animals in the first place: reason. And acting without reason—imagining the position of the animal to be like the position of the self—is acting only with what is shared by the human and animal: the sensitive rather than the rational soul. To identify with an animal from the perspective of this theoretical framework is to limit oneself to the being of an animal, and thus a transformation has taken place.

But in staging Jaques' compassion as unreasonable, Shakespeare is not merely deriding such a feeling. He is showing the limits of compassion in orthodox thought—it excludes animals—and revealing its failure to deal with something that is both very real and strangely recognizable: the "sobbing deer." In fact, Jaques' feeling for the stag is not completely aberrant at the end of the sixteenth century. Not only is the image of the weeping deer one that has a classical pedigree and is repeated in a number of works of the period;[47] it also links Jaques with an alternative ethical framework that was

---

[44] Shakespeare, *As You Like It*, 2.1.27–28, 2.1.58–63.

[45] The idea of "imaginative dwelling" comes from Nancy E. Snow, "Compassion," *American Philosophical Quarterly* 28, no. 3 (1991): 195, 198.

[46] Shakespeare, *As You Like It*, 2.7.1–2.

[47] See, for example, Ovid, *The XV. Bookes of P. Ouidius Naso, entytuled Metamorphosis, translated oute of Latin into English meeter, by Arthur Golding Gentleman* (London: William Seres, 1567), Book III, p. 63, for the tears of the transformed Acteon; Stephen Bateman, *Batman vppon Bartholome, his Booke De Proprietatibus Rerum* (London: Thomas East, 1582), fol. 358r; Robert Chamberlain, *Nocturnall Lucubrations: Or Meditations Divine and Morall* (London: M. F. for Daniel Frere, 1638), 30. The hart was also said to weep each year when it lost its antlers: see Guillaume de Salluste Du Bartas, *Bartas His Deuine Weekes & Workes*, trans. Joshuah Sylvester

emerging at that time. And this new ethical framework offers a different vision of the human in part because it returns the focus to real rather than symbolic animals. We should, I think, turn to the ideas of Michel de Montaigne in order to understand more fully Jaques' status in *As You Like It*.

The affiliation of Jaques with Montaigne can be traced through a powerful image that is repeated by both. Jaques' compassion for the stag echoes Montaigne's own: the latter writes that when in the hunt "the stag, feeling himself out of breath and strength, having no other remedy left, throws himself back and surrenders to ourselves who are pursuing him, asking for our mercy by his tears," this is "a very unpleasant spectacle."[48] Once again, it is through an animal, this time a tearful stag, that we can begin to understand a way of thinking about the nature of the human and the significance of reason in early modern thought.

Montaigne's essay in which the stag's tears appear is titled "Of cruelty" and has been read by David Quint as the "central statement of [Montaigne's] moral position."[49] The essay was first published in 1580 and was expanded as Montaigne returned to his *Essais* between 1580 and 1588. It is, so Hallie argues, "one of the most powerful essays on ethics ever written. . . . In a few pages it manages to explore and explode one of the main traditions in the history of man's thought about good and evil, and then—again with remarkable brevity—it makes a statement about ethics that illuminates and gives vitality to the usually heartless abstractions of Western ethics."[50] What Montaigne does that is so remarkable at that date is turn away from the self that is central to inward government theory and look instead at the other, at the individual on whom cruelty is inflicted.

The essay begins with a discussion of virtue in which Montaigne sets up a distinction between "goodness," which he argues is a natural predisposition, and "virtue," which is "more active." There is a difference between someone who despises "injuries received" through a "natural mildness and easygoingness" and another "who, outraged and stung to the quick by an injury should arm himself with the arms of reason against this furious appetite for vengeance, and after a great conflict should finally master it." The display of virtue, Montaigne argues, requires "opposition," and this is a reason why God can be described as good, but not virtuous: "his operations are wholly natural and effortless."[51] However, Montaigne goes on, not only is

---

(London: H. Lownes, 1605–6), 195. See also Steven Doloff, "Jaques' 'Weeping' and Ovid's Cyparissus," *Notes and Queries* 239 (December 1994): 487–88.

[48] Michel de Montaigne, "Of cruelty," in *The Complete Works*, trans. Donald M. Frame (1948; repr., London: Everyman, 2003), 383.

[49] David Quint, "Letting Oneself Go: 'Of Anger' and Montaigne's Ethical Reflections," *Philosophy and Literature* 24, no. 1 (2000): 126.

[50] Hallie, "Ethics of 'De la cruauté,'" 156.

[51] Montaigne, "Of cruelty," 372.

what is perceived as virtuous sometimes accidental—"want of apprehension and stupidity sometimes counterfeit valorous actions"—but as well as this the distinction between goodness and virtue leads to a question that offers an important prospect for thinking about the nature of cruelty: "If," he asks, "virtue can shine only by clashing with opposing appetites, shall we then say that it cannot do without the assistance of vice, and that it owes to vice its repute and its honor?"[52] To be virtuous, one must also be potentially vicious; this is the logic of the discourse of reason.

For Montaigne, however, the virtuous struggle to restore order, reason's attempt to control vicious impulses, is, in Hallie's terms, "an irrelevancy."[53] For Montaigne, this abandonment of the focal point of the discourse of reason is logical. Virtue, he argues, is far from simply evidenced in self-control. In fact, at the very center of the essay he confesses what he terms "a monstrous thing" about himself: "I find . . . in many things . . . my lust less depraved than my reason."[54] The Stoic denigration of passion, the desire for apathy (a lack of passion), is revealed by Montaigne as immoral, but this is not to reiterate the attack on Stoicism made by Wright or Hall. As Hallie puts it, for Montaigne, "the non-philosophic, the instinctual (or at least the second-natural or habitual) powers in a person can be safer guides to conduct than unbridled reasonings."[55] Sometimes acting without reason can be more genuinely good than acting with reason.

But as well as challenging the centrality of reason, "Of cruelty" also offers a new pivotal category to ethics. Montaigne writes of witnessing executions, and states, "Savages do not shock me as much by roasting and eating the bodies of the dead as do those who torment them and persecute them living. Even the executions of the law however reasonable they may be, I cannot witness with a steady gaze." He goes on: "all that goes beyond plain death seems to me pure cruelty."[56] For Montaigne, cruelty, even in lawful executions, goes beyond what is necessary for the eradication of a social threat, and therefore torture, he argues, should be exercised on the dead bodies of criminals, not on the living: "against the shell, not the living core." This torture of the dead would, he writes, "affect the common people" just as would an excessive public execution, but it would maintain the crucial difference between the sentient (who should be protected from cruelty) and the insentient (the dead, who require no such protection).[57]

This undermining of the supremacy of reason and the emphasis on sentience inevitably leads Montaigne to think about animals. And instead of

---

[52] Ibid., 376, 374.
[53] Hallie, "Ethics of 'De la cruauté,'" 171.
[54] Montaigne, "Of cruelty," 378.
[55] Hallie, "Ethics of 'De la cruauté,'" 164.
[56] Montaigne, "Of cruelty," 380–81.
[57] Ibid., 382.

maintaining a focus on the (human) self, Montaigne treats animals as moral patients, as beings that suffer: "I have not even been able without distress," he writes, "to see pursued and killed an innocent animal which is defenseless and which does us no harm." At another point in the essay he writes, "I do not see a chicken's neck wrung without distress, and I cannot bear to hear the scream of a hare in the teeth of my dogs, although the chase is a violent pleasure."[58] Just as virtue requires vice, so violence meets pleasure; cruelty meets enjoyment. There is no sense in which virtue can, with the work of reason, overcome and dismiss vice; there is instead, inevitably for Montaigne, an uncomfortable coexistence.

The reason for Montaigne's distress at the suffering of animals, then, is not merely to be regarded as "a point of softness" (his words); he is not simply sentimental. Rather, his distress is logical. Montaigne writes of animals: "There is some relationship between them and us, and some mutual obligation." Animals, unlike the dead bodies of humans, are sentient, and can, if only by basic means, communicate their suffering.[59] In another essay Montaigne refers to Pyrrho's pig, who, according to Diogenes Laertius, remained calm on a ship caught up in a storm while the human passengers were "unnerved."[60] Montaigne states, "To be sure, he is unafraid in the presence of death, but if you beat him he will squeal and wiggle."[61] In Montaigne's mind the pig may well illustrate admirable calmness in the face of unchangeable fate, but this is not the point; it is the pig's ability to feel pain that is important. The discourse of reason does not, of course, deny sentience to animals—humans share with animals many sensory capacities—but what it does do that Montaigne refuses to follow is regard sentience as the less important quality. When reason is the central concern, it is possible, as has been shown, to deny the reasonable status of the other in order to make it the victim of cruelty. For Montaigne, however, the basis of ethics is a perception of shared sentience, and this denies the power of dehumanization, allowing individuals to recognize themselves in the other. And from this recognition, he argues, comes society, fellow feeling. This is crucial. Montaigne writes, "I sympathize very tenderly with the afflictions of others."[62] This sympathy has, in Hallie's terms, taken "ethical thought out of its egocentricity, out of its exclusive concern for the moral government of the Moral Agent, and into the world in which the Moral Agent is a part."[63]

---

[58] Ibid., 383, 379.

[59] Ibid., 379, 385.

[60] Diogenes Laertius, *Lives of Eminent Philosophers,* trans. R. D. Hicks (London: William Heinemann, 1925), 2:481.

[61] Montaigne, "That the taste of good and evil depends in large part on the opinion we have of them," in *Complete Works,* 44.

[62] Montaigne, "Of cruelty," 380.

[63] Hallie, "Ethics of 'De la cruauté,'" 166.

Whereas in the discourse of reason the death of an animal could be as insignificant an ethical concern as the theft of a cart because of the emphasis on reason (the ethical concerns would be centered on the economic loss suffered by the owner), in Montaigne's thought, loss to the owner of the animal does not constitute the focus; it is the suffering of a sentient being that counts. The possibility of a relationship and of mutual obligation is breeched in an act of cruelty, and the ideal of friendship that Montaigne discusses in the essay of that title becomes impossible. Thus, cruelty, he says, is "the extreme of all vices."[64]

Prudence is the super-virtue in inward government theory: it uses the human's capacity to recall the past, contemplate and understand the present, and plan for the future; it is, in short, the essence of reason in the discourse of reason. David Quint has argued that "On cruelty," however, is "an attack on prudential wisdom," an attack on "the idea that one can use historical analogy to determine the proper course of present behavior."[65] Prudence is dismissed because it is the present, the real and experienced, that is important when writing of cruelty. It is at the moment when the stag weeps that the relationship should be formed. Cruelty, for Montaigne, ignores sentience, uproots friendship, and is thus the incarnation of violence in a world in which it is the relationship between self and other that is truly constitutive. Cruelty is, in short, the super-vice. And it is with some terrible irony that he can term cruelty an "ordinary" vice, for in its ordinariness cruelty reveals the customary essence of the human to be not reasonable—far from it—but mundanely violent and uncaring. And it should come as no surprise in this context that for Montaigne good memory, so central to the discourse of reason, merely makes humans good liars.[66]

In Montaigne's work, then, there is a turning away from assertions of human superiority and the significance of the rule of reason that is rare in this period. But his inclusion of animals as animals, rather than as objects, within the human moral framework can be found in the work of other writers in England. Whereas Shakespeare represented Montaigne's ethics in order to parody them, other writers take them more seriously. What perhaps links Montaigne to these English writers is not nationality or religion—the works that follow are by English Protestants, while Montaigne was a French Catholic—but the sense in which it is real animals that are the focus. Montaigne speaks of the stag's tears in a way that conveys a moment of actual fellowship, and in the work of these English Protestants we can trace a similar shift from the notion of the animal as a symbol to a real being in the world.

---

[64] Montaigne, "Of cruelty," 379.

[65] David Quint, *Montaigne and the Quality of Mercy: Ethical and Political Themes in the Essais* (Princeton, N.J.: Princeton University Press, 1998), 20.

[66] Montaigne, "Of cannibals," and "Of liars," in *Complete Works*, 189 and 27.

Feeling Subjects

Some of the shifts in focus from the ethics of inward government to the possibility of animal subjectivity are simple. In his 1625 discussion of Balaam
and the ass (Numbers 22:21–33), Joseph Hall, for example, begins with the
miracle of the speaking ass "whose common sense is aduanced aboue the
reason of his rider," and argues that this is an example of the power of
the Almighty: "There is no mouth, into which God cannot put words: and
how oft doth hee choose the weake, and vnwise, to confound the learned,
and mighty." This theological discussion, however, leads to a practical assertion. Hall writes:

> I heare the Angell of God taking notice of the cruelty of *Balaam* to his beast:
> His first words to the vnmercifull prophet, are in expostulating of this wrong.
> We little thinke it; but God shall call vs to an account for the vnkind and cru-
> ell vsages of his poore mute creatures: He hath made vs Lords, not tyrants;
> owners, not tormenters.[67]

Here cruelty to animals is something that is not a path to sin (Aquinas's view)
but is sinful in and of itself. Animals, not the owners of the animals, are worthy recipients of kind acts by moral agents. In this, Hall has moved the
boundaries of community, has included animals within his moral framework. From a modern perspective, such a shift might not seem huge, and
the paradoxes that persist in Montaigne's ethics (he goes hunting, he eats
meat but he believes in a communication between human and animal) can
also be traced in Hall's work when he goes on to say that "hee that hath giuen
vs leaue to kill them, for our vse, hath not giuen vs leaue to abuse them, at
our pleasure; they are so our drudges, that they are our fellowes by creation."[68] This seems to return to Aquinas's sense that animals are on earth
to serve humans, but Hall is making an important distinction. While it is ac-
ceptable to kill animals for use, they are not to be the victims of our recreation. Animals, in Hall's representation are "our fellows by creation": they
share our world. Even though they have a lower place than humans, they
still have a place.

Hall had already outlined this view in his 1607 *Mediations and Vowes Diuine
and Morall*. Here he presented a vision of the chain of being, but recognized,
crucially, that humans and animals had a closeness that should have alerted
the former to their responsibility toward the latter. He wrote, "I doe there-

---

[67] Joseph Hall, *Contemplation, Lib. VII, "Of Balaam,"* in *The Works of Joseph Hall Doctor in
Diuinitie, and Deane of Worcester* (London: Thomas Pauier, 1625), 934–35.
[68] Ibid., 935.

fore owe awe vnto God; mercie to the inferiour creatures: knowing, that they are my fellowes, in respecte of Creation; whereas there is no proportion betwixt me, and my Maker."[69] This is significantly different from the anthropocentric view that presented animals as mere objects, a view which regarded the killing of animals as robbery rather than homicide and which regarded cruelty to animals as detrimental to human salvation but not to the animals themselves.

Likewise, the Jacobean writer John Bruen moves from the conventional Thomist position to take in more fully the similarities of humans and animals. He writes, "Mee thinkes these gentlemen's horses being so grosly abused should likewise rebuke the fiercnes and foolishnes of their masters, if not by mans voice, yet by the voices of their grievous grones which they heave from them, when being over-rid, past their strength and breath." The sound of the horse is not a mere sound; rather it is, like the tears in the eyes of the stag, to be interpreted as meaningful and as conveying a real sense of injustice. Bruen draws an analogy that reiterates Hall's distinction between useful and pleasurable killing, and he emphasizes fellow feeling and the link between children and animals:

> A good rule for our horse-racers, rank riders, and hot-spurre hunters (if they have grace to follow it) in all their recreations and pursuits of their pleasures, to measure their actions and moderate their passions by; that as they may and ought to have a care to charge no burden upon their children but such as they may well beare, so they may not over draw, nor over-drive their beasts for one day, nor put them to any toyle and travell, but that which they are able to indure.[70]

If children are like animals, then, logically, animals are like children, and recognizing this inversion offers a space for an ethical treatment of animals that takes their sentience seriously.

An even more explicit statement of the significance of sentience can be traced in Robert Cleaver's *A Plaine and Familiar Exposition of the Eleuenth and Twelfth Chapters of the Prouerbes of Salomon* (1608). Here Cleaver begins to show how a move from the discourse of reason to the notion of fellowship might manifest itself in ethical discussion. Encountering Proverbs 12.10 once again, we read:

[69] Joseph Hall, *Meditations and Vowes Diuine and Morall* (London: Humfrey Lownes, 1607), 76.

[70] *The Life of John Bruen of Bruen Stapelford,* cited in *The Journal of Nicholas Assheton of Downham, in the county of Lancashire esq.,* ed. F. R. Raines (Chetham Society, 1848), 14:27.

Mercy is to be shewed not onely to men, but to the vnreasonable creatures also. As all creatures doe taste of, and liue by the aboundant liberality and bountifulnes of Gods hand, so would he haue them to feele by sense, though they cannot discerne it by reason, that there is also care for them and compassion in his children.[71]

Here animals' lack of reason is regarded as a lack (in this Cleaver is different from Montaigne), but that lack is not all that is considered. Instead, and more like Montaigne, Cleaver asserts animals' ability to feel as the more important ethical point. It is for this reason that humans are to show mercy to them. In acknowledging the sentient nature of animals—their God-given capacity to feel in the world—Cleaver shifts his ethics to allow for this fact. Reason is not all that is worth recognizing, and if this is the basis of ethical discussion, it is inevitable that humans can be understood to have fellowship with animals, that there is a kind of equality between them that could never exist in the discourse of reason.

We have come a long way from the "preferred reading" of the discourse of reason with which I started here. That discourse begins with an opposition between real humans and real animals (the failure to trace the rational soul in anatomical examination does not make the soul any less real). But as that opposition fails to hold up—as humans *become* animals—so real animals are driven from the discourse. This disappearance of real animals serves an important function. It is as if humans, as they abandon the use of the thing that makes them human and become, by logic, animals, can only be safely comprehended if those animals cease to be, to put it simply, animals; because if humans really do become animals, or even worse than animals, then the category "human" would be virtually meaningless. In order, I think, to defend against this logic, early modern writers shift from representing the decline of the human as real (and logical) to treating it as metaphorical. A human becomes *like* an animal, but does not actually become an animal. However, the metaphorical representation of the decline, we should remember, works against the logic of the human that is established under an equivalence of binaries: human is to animal as reasonable is to unreasonable. In fact, where these binaries help to establish the human, they also begin to undo it. This undoing of the binary was actually emphasized in an alternative ethic, one that not only took cruelty as a central danger to human status but also recognized that the achievement of human status required kindness to animals, not because animals were objects of human superiority but because they

---

[71] Robert Cleaver, *A Plaine and Familiar Exposition of the Eleventh and Twelfth Chapters of the Proverbes of Salomon* (London: Richard Bradocke, 1608), 140–41.

were fellow beings. Human as a category thereby ceased to have the meanings that were attached to it in the discourse of reason because, in part, that discourse was perceived to be logically flawed.

However, that is not to say that the flaw in the discourse was the only reason for the emergence of this new ethical framework. This cannot be the case because the discourse that was so central in the early modern period is one that first emerged—with all of its flaws and inconsistencies—in classical writings centuries earlier. What was important in early modern English thought was that these flaws became a focus at a time when alternative ancient ideas about humans and about animals were re-entering discussions. These ideas did not merely uproot the theory that set human in opposition to animal in relation to the possession of reason; they emerged out of that discourse and responded to its failings. And these ideas can be seen in motion most clearly in early modern discussions of animals. Here we are still to a great extent dealing with discussions that use animals as a way of explaining the nature of the human; but what also emerges in these explanations is further discussion of the nature of the animal itself.

CHAPTER 4

Being Animal

If the human was constructed and undone in the discourse of reason and if the animal was frequently used as the other of the human to reinforce and—paradoxically—to undercut the superiority of that human, then it should come as no surprise to find that the discourse of reason offered designations of *animals* that replicated the defining structures of the human. In these structures humans were perceived to differ absolutely from animals—they had reason while animals lacked it—and the emphasis that was placed on education and on actions—what humans actually did—meant that the simple a priori opposition between human and animal became, to say the least, more complex. As well as this, and following the logic of the discourse of reason, on one level humans had a capacity that meant that they could be virtuous, but on another level this capacity also meant that humans could be vicious in a way that animals never could. Thus the discourse of reason itself both made and undid human superiority.

A similar trajectory can be traced in the ways in which animals are defined and discussed in the discourse of reason. To some extent, that definition has already been outlined: animals were absolutely other; they lacked reason and therefore lacked access to the past and the future, living only in the present. But early modern writers had more to say about animals than this: moving outside of the discourse of reason, the perception of animals' natural knowledge played a role in undoing the opposition. Once again, from a rather different perspective, the notion of animal superiority over the human emerges.

This chapter traces the difference between what I am terming theoretical and empirical assessments of animals; between ideal and real beasts.

(The distinction is often hard to see because *ideal* is a term that can be applied both to animals of flesh and blood and to beasts of the imagination.) The chapter also shows how the conceptual shifts within the definition of the human in the discourse of reason are replicated in discussions of animals. By tracing such a movement in early modern ideas about animals, it is possible to see how empirical observation begins to undercut assertions of animal irrationality; how, just as looking at human behavior can challenge the seemingly parallel binaries of human / animal and reason / unreason, so looking at animal behavior can also uproot difference. It does not seem to be a coincidence that, at the moment this complexity was flourishing in early modern English writing, a classical school of thought was rediscovered which offered a very different view of animals. This chapter does not conclude with an assertion of fellow feeling and the inclusion of animals within human ethical debates (one end of the discussion of the human). Rather, it ends with skepticism, with the abandonment of human knowledge in the face of the possibility that a dog might know more than we can know that it knows. But the starting point of this chapter is not the undoing of structures of thought so much as their inversion.

Organic Reason

The notion of the natural superiority of animals that we have already encountered—their lack of vice—was a logical outcome of the discourse of reason. But an alternative way of asserting animals' superiority offers a different challenge to the Aristotelianism that lies at the heart of that discourse: animal instinct, this alternative tradition argues, could set animal above human. One key source of this construction of the natural order was Plutarch. Roger French has argued that Plutarch (46–120 CE) followed Aristotle in "seeing the essence or nature of the animal as its soul, the character of which determines the nature of the physical body and the behaviour of the animal."[1] Where Plutarch differs from Aristotle, however, is in his argument about the nature of the animal's soul. Plutarch, in fact, takes the foundation of his ideas not from Aristotle but from Plato, and this opens up possibilities for understanding the nature of animals that the discourse of reason cannot allow for.

The discourse of reason, as I have outlined it so far, took its foundation from Aristotle's theory of the three souls: vegetative, sensitive, and rational. However, existing alongside this theory in the early modern period, and sometimes confusing the clarity of this theory (remember Henry King's em-

[1] Roger French, *Ancient Natural History: Histories of Nature* (London: Routledge, 1994), 179.

phasis on the *organ* of memory), another model of the human persisted, a model that came from the theories that Aristotle moved away from. For us, the crucial distinction between Aristotle's and Plato's views is Plato's emphasis on what might be termed by Aristotelians the organic nature of reason. In *Timaeus* Plato argues that reason is not to be found, as Aristotle would propose, in an immaterial essence; rather, Plato suggests that it is placed in "the divinest part" of the human's body. The body, in fact, is represented as a "convenient vehicle" for reason, which is housed in the brain and which is separated from the dangerously earthly nature of the body by the neck, which is "a kind of isthmus and boundary between head and breast to keep them apart." The vertical nature of the body is used as an image of its capacity: the head on the top signifies the (ideal) predominance of reason, below which are the emotions (housed in the heart). The midriff separates the heart from the belly, and it is in the belly that the appetites are placed.[2]

Plato's model of the human is given its finest outline in the early modern period in Book II of Edmund Spenser's *Faerie Queene* (1590). Here, in Canto IX, Guyon and Arthur arrive at the Castle of Alma, the "house of Temperance," and their journey into and through the castle reveals it to be a Platonic model of the human. Having entered the castle, the knights pass first through "a stately Hall / Wherein were many tables faire dispred," where they encounter the "Marshall of the same, / Whose name was *Appetite*." They continue past some "bellowes, which did styre / Continually, and cooling breath inspyre," and see the "kitchin Clerke, that hight *Digestion*" who sends the "fowle and wast" matter "to the back-gate" where it is "auoided quite, and throwne out priuily." The knights are then led into another chamber, "a goodly Parlour," where ladies sit with Cupid, and where with "Diuerse delights they found them selues to please." They continue their journey, led by Alma, "Vp to a stately Turret:"

> The roofe hereof was arched ouer head,
> And deckt with flowers and herbars daintily;
> Two goodly Beacons, set in watches stead,
> Therein gaue light, and flam'd continually:
> For they of liuing fire most subtilly
> Were made, and set in siluer sockets bright,
> Couer'd with lids deuiz'd of substance sly,
> That readily they shut and open might.

Within the turret the knights meet "three honorable sages"; "The first of them could things to come foresee: / The next could of things past best

---

[2] Plato, *Timaeus*, trans. Desmond Lee (Harmondsworth: Penguin, 1971), 61, 97.

aduize; / The third things past could keepe in memoree." In the first chamber is "*Phantastes*"; the second sage in the middle chamber is unnamed, but meditates on "All artes, all science, all Philosophy, / And all that in the world was aye thought wittily." In the final, "hindmost" chamber sits "an old oldman, halfe blind." "This man of infinite remembrance was," aided by "a litle boy," who

> did on him still attend,
> To reach, when euer he for ought did send;
> And oft when things were lost, or laid amis,
> That boy them sought, and vnto him did lend.[3]

Here, Spenser has created a Platonic image of the human, with the journey from appetite to emotion to reason represented as a journey not from the organic to the inorganic but from belly to heart to brain. Whereas Aristotle and, following him, the writers we have already encountered regarded the brain as having merely organic, sensitive capacities—the limitation of the animal—in this Platonic model the brain is the seat of reason. This, inevitably, opens up the possibility that, because animals have brains, animals also have the capacity to reason. Plutarch, following this line of reasoning, longer perceived animals as lesser, missing the crucial third—rational—soul: as Roger French puts it, in Plutarch's ideas animals "were people."[4] But French might have taken this summary even further: for Plutarch, it might be fairer to say, animals were *good* people.

Human Inferiority

Plutarch argued that animals are reasonable and that they are more virtuous than humans. The first claim can be found in his dialogue entitled, in the 1603 English translation, *Whether Creatures Be More Wise, They of the Land, or Those of the Water.* Here Plutarch's speakers begin with a complaint about hunting and move to a debate on the status of the wisdom of animals and sea creatures (thus linking animals' ethical standing with their rational capacity). The speakers start with a simple proposition: that "all sorts of living creatures have in them some little discourse and reason." Such a statement,

---

[3] Edmund Spenser, *The Faerie Queene* (1596), in *The Poetical Works of Edmund Spenser* ed. J. C. Smith and E. De Selincourt (Oxford: Oxford University Press, 1924), Book II, Canto IX, passim. Michael C. Schoenfeldt discusses this passage of *The Faerie Queene* in *Bodies and Selves in Early Modern England: Physiology and Inwardness in Spenser, Shakespeare, Herbert, and Milton* (Cambridge: Cambridge University Press, 1999), 40–73.

[4] French, *Ancient Natural History*, 179.

one named Autobulus argues, is logical: "necessarie it is, that everie creature which hath sense, should likewise be endued with discourse of reason and understanding." Unlike in Aristotelianism, where sense and reason were regarded as separate faculties (one of the body, the other immaterial), in this Platonist's thought, sense is inseparable from reason as they are both situated in the body. Autobulus goes on: "if [animals'] witte be more dull than ours, it followeth not thereupon, that they have neither reason nor naturall witte." In illustration he tells numerous stories of apparent animal wisdom: of "their provisions and forecasts, their memories, their affections, their tender care of their yong ones, their thankfulnesse to those who have done thē good, their hatred & rankor against them who have done them a shrewd turne: their industry to find out things necessary for them." Not to be outdone, Phædimus, another of Plutarch's interlocutors, likewise proclaims the wisdom of "sea creatures" which

> generally all, have a certeine in bread sagacity, a wary perceivance before hand, which maketh them to be suspicious and circumspect, yea, and to stand upon their guard against all fore-laying; so that the arte of hunting and catching them is not a small piece of worke, and a simple cunning; but that which requireth a great number of engines of all sorts, and asketh wonderfull devices.

It is the difficulty of asserting dominion that offers evidence of animal reason.[5]

In another text, however, Plutarch goes even further than this. *That Brute Beasts Have Use of Reason,* more commonly known as *Gryllus,* is a dialogue between Circe, Ulysses, and Gryllus, a human who has been transformed by Circe into a pig. Told first in Homer's *Odyssey* and then in Ovid's *Metamorphoses,* the story of Circe's transformation of Ulysses' followers was, as has already been shown, popular in the early modern period as an image of human decline into passion and vice. In Arthur Golding's translation of *The Metamorphoses* Achemenides, one of Ulysses' followers, relates his own transformation and repeats Plato's idea of the upright—rational—human:

> I wext all rough with bristled heare,
> And could not make complaint with woordes. In stead of speech I there
> Did make a rawghtish grunting, and with groueling face gan beare
> My visage downeward too the ground.

---

[5] Plutarch, *Whether Creatvres Be More Wise, They of the Land, or Those of the Water,* in *The Philosophie, commonlie called, The Morals Written by the learned Philosopher Plutarch of Chærnea,* trans. Philemon Holland (London: Arnold Hatfield, 1603), 951, 952, 955, 958, 970.

Ultimately the return to human form is greeted with joy. Golding's Ovid writes,

> The more shee charmd, the more
> Arose wee vpward from the ground on which wee daarde before.
> Our bristles fell away, the clift our clouen clees forsooke.
> Our shoulders did returne agein: and next our elbowes tooke
> Our armes and handes theyr former place. Then weeping wee enbrace
> Our Lord, and hung about his necke whoo also wept apace.
> And nat a woord wee rather spake than such as myght appeere
> From harts most thankfull too proceede.[6]

In Plutarch's *Gryllus*, however, the story is very different. Here Ulysses asks Circe to introduce him to any of his fellow humans who have been transformed into animals. Ulysses follows an argument central to the foundation of the discourse of reason, proposing that there is a crucial (and naturally—although clearly not supernaturally—unbridgeable) gap between humans and animals, and he tells Circe that he wishes to "receive them men againe, and save them, strangers though they be." Human status is to be returned in order that the transformed men's "suffer[ing] against nature" should cease. Circe responds to Ulysses' accusation of the unnaturalness of the transformations with a suggestion that is nowhere to be found in Homer or Ovid: "See the simplicitie of this man; he would through his folly, that his ambitious minde should procure damage and calamity not to himselfe onely and his friends, but also to those who are meere aliens, and nothing belonging to him?" To return these beasts to human form is, in her opinion, to damage them, not to rescue them.

Unable to comprehend Circe's disdain, Ulysses continues with his zealous—and Aristotelian—humanizing, and attempts to persuade Gryllus, now a pig, to return to human shape. Gryllus's response is clear:

> you would perswade us, that whereas we live not in abundance, and enjoy the affluence of all good things, we should quit the same, and withall, abandon and forsake her who hath procured us this happinesse, and all to goe away with you, when we are become men againe; that is to say, the most wretched creatures in the world.[7]

Gryllus wishes to remain in a pig's body, and Ulysses, in conventional fashion, accuses Gryllus of being drunk on Circe's potion which has "spoiled

---

[6] Ovid, *The XV. Bookes of P. Ouidius Naso, entytuled Metamorphosis, translated oute of Latin into English meeter, by Arthur Golding Gentleman* (London: William Seres, 1567), Book xviiij, sigs. 177^{r-v}.

[7] Plutarch, *That Brute Beasts Have Use of Reason*, 562, 564.

[his] wit and understanding." Gryllus answers with a paean to animals: "the soule of brute beasts," he says, "is by nature more kinde, more perfect and better disposed to yeeld vertue, considering that without compulsion, without commandement, or any teaching, which is as much to say, as without tillage and sowing it bringeth forth and nourisheth that vertue which is meet and convenient for every one." Humans must learn to be civilized, to be human; animals have their civilization (and, by implication, paradoxically, their human status) naturally.

What follows Gryllus's statement are examples of "virtuous" animal behavior: animals are clean fighters and lack the "deceit" that men use; they would rather refuse food and starve to death than "live in servitude"; they are "more continent" than men. And while nature, Gryllus argues, "is not able to keepe [men's] intemperance within the limits and bounds of reason" and to stop them from having sex with "shee goats, with sowes and mares, . . . there was never found yet any brute beast to have lusted after man or woman."[8] To be an animal, in this text, is to be more natural and less vicious than, and thus superior to, a human. As Philemon Holland notes in his "Summarie": Gryllus *sheweth sufficiently, that if men have no other approch to rest upon, than a naturall habitude of an earthly vertue, and can assure the repose of their consciences upon nothing but upon humane valliance, temperance, and wisedome, they doe but goe in the companie of beasts, or rather come behind them."*[9] Animals have virtue naturally, while humans are dangerously vicious; whereas animals fulfill their domestic duties without deliberation, humans must be schooled to complete theirs. The fact that being human must be taught, while being animal is natural, is taken not as evidence of the power of the human potential to reason but as evidence of human frailty. Gryllus cites the example of animals' ability to treat their own illnesses and asks, "who hath taught the tortoises, when they have eaten a viper, to seeke out the herbe *Organ* for to feed upon?" He answers his own question immediately:

> if you say (as the trueth it is) that nature is the schoole-mistresse, teaching them all this, you referre and reduce the wisdome and intelligence of dumbe beasts unto the sagest and most perfect cause or principle that is; which if you thinke you may not call reason, nor prudence, ye ought then to seeke out some other name for it, that is better and more admirable, as being neither ignorant nor ill taught, but having learned rather of it self, not by im-

---

[8] Ibid., 568. Aelian (c. 170 CE) offers a different opinion of this, telling tales of a dog who fell in love with "Glauce the harpist," and a seal who fell in love with a diver. Aelian, *On the Characteristics of Animals*, trans. A. F. Scholfield (London: Heinemann, 1958), 21, 279. On this see Joyce E. Salisbury, *The Beast Within: Animals in the Middle Ages* (New York: Routledge, 1994), 84–88.

[9] Philemon Holland, "Summarie" to Plutarch, *That Brute Beasts Have Use of Reason*, 562.

becilitie and feeblenesse of nature, but contrariwise, through the force and perfection of naturall vertue, letting go, and nothing at all esteeming that beggerly prudence which is gotten from other by way of apprentissage.[10]

For humans, education makes up for a "feeblenesse of nature," it is a necessary addition to a fragile being. Animals, in contrast, are taught by nature, or rather need no teaching because what they know, they know naturally.

But more threatening to notions of human superiority than this is the sense, not that education fulfills human potential (thus proving it to exist in the first place) but that it makes men servile: "And thereupon I inferre and conclude," says Gryllus to Ulysses, "that you and such as you are, exercise a kind of valiance (I must needs say) which is not voluntarie nor naturall, but constreined by force of lawes, subject and servile to (I wot not what) customes reprehensions."[11] Animals are not, as many thinkers had argued, slaves to instinct; rather, it is humans who are slaves, slaves of their customs, of their civilization. Free will, in this context, it seems, is a myth of the discourse of reason, is a legend used by humans to bolster their own status. Thus Plutarch's dialogue between Gryllus and Ulysses can be read more generally as a debate between Plutarchianism and Aristotelianism in which Ulysses the Aristotelian is presented as being confined as much by his discourse as by his humanity.

Plutarch's version of the Circe myth emerges in other places in early modern English culture, and sometimes with very different interpretations. In *A Choice of Emblemes* (1586), for example, Geffrey Whitney includes the following: "*Homines voluptatibus transformantur*" (see fig. 4.1). And while Whitney lists "Ouid. Metam. lib.14" in the margin, the version of the Circe myth he relates comes from Plutarch, even if the moral he draws does not:

See here VLISSES men, transformed straunge to heare:
Some had the shape of Goates, and Hogges, some Apes, and Asses weare.
Who, when they might haue had their former shape againe,
They did refuse, and rather wish'd, still brutish to remaine.
Which showes those foolishe sorte, whome wicked loue dothe thrall,
Like brutishe beastes do passe theire time, and haue no sence at all.
And thoughe that wisedome woulde, they shoulde againe retire,
Yet, they had rather CIRCES serue, and burne in theire desire.
Then, loue the onelie crosse, that clogges the worlde with care,
Oh stoppe your eares, and shutte your eies, of CIRCES cuppes beware.[12]

---

[10] Plutarch, *That Brute Beasts Have Use of Reason*, 569.
[11] Ibid., 565.
[12] Geffrey Whitney, *A Choice of Emblemes* (Leyden: Christopher Plantyn, 1586), 82.

Figure 4.1. *"Homines voluptatibus transformantur,"* from Geffrey Whitney, *A Choice of Emblemes* (Leyden: Christopher Plantyn, 1586), p. 82. By permission of the British Library, classmark G11572.

Similarly, at the end of Book II of *The Faerie Queene* after the Bower of Bliss has been destroyed by the rampaging Guyon, the Palmer explains "what meant those beastes, which there did ly."

> Said he, These seeming beasts are men indeed,
> Whom this Enchauntresse hath transformed thus,
> Whylome her louers, which her lusts did feed,
> Now turned into figures hideous,

> According to their mindes like monstruous.
> Sad end (quoth he) of life intemperate,
> And mournefull meed of ioyes delicious.

Guyon—like Ulysses—asks for these beasts to be returned to human form. The Palmer agrees, but one among them "Repined greatly" and complained about his return to human form; his name, Grill.

> Said *Guyon*, See the mind of beastly man,
> That hath so soone forgot the excellence
> Of his creation, when he life began,
> That now he chooseth, with vile difference,
> To be a beast, and lacke intelligence.
> To whom the Palmer thus, The donghill kind
> Delights in filth and foule incontinence:
> Let *Grill* be *Grill*, and haue his hoggish mind,
> But let vs hence depart, whilest wether serues and wind.[13]

As Michael C. Schoenfeldt has noted, in Spenser's version of this story Grill "represents essential, unregulated, and unreconstructed humanity."[14] It is a far cry from the Plutarchian vision of Gryllus as the spokesman for a superior nature.

It is interesting to note, in the light of this orthodox avoidance of Plutarch's version of the Circe myth, that his ideas can be traced, if not in fiction, then in natural philosophy. And even though few thinkers take up Plutarch's whole theory with any thoroughness, versions of his vision of animals as the natural possessors of virtue can be traced throughout the early modern period. While in some—orthodox, that is, Aristotelian—texts the goodness of animals is based on their lack of reason (and therefore on their natural lack of vice), in others—those that follow Plutarch more closely—it is animals' possession of a *natural* reason that makes them superior to humans. When, in 1561, Pierre Viret suggested, for example, that we attend the *"Schoole of Beastes,"* he was making an argument that reversed the trajectory that was in place in orthodox discussions of children. There, schooling was seen as a way of undoing the natural beastliness of humanity, of turning children into humans. In Viret's text, however, humans are sent to school not in order to leave the beasts behind but in order to learn from them. As the author notes: *"I Haue intituled this Dialogue, the good Householder, or the* Oeconomicks, *because I make comparison in the same, of the good and euil householders*

---

[13] Spenser, *Faerie Queene*, Book II, Canto XII, stanzas 85–87.
[14] Schoenfeldt, *Bodies and Selves*, 71.

with the beastes, which knowe best to prouide for their nourishment and conseruation, as well of them as of their yong." The source for many of the examples given and for the overriding philosophical argument that Viret makes is clear: as his authority for the statement that "al the beastes generally . . . do loue ētierly that which they ingender and bring forthe, and cherish them carefully" he writes, "as *Plutarck* witnesseth."[15] But Viret is not alone in his use of Plutarch, and we can look first to some of his fellow countrymen to see the fullest early modern expressions of this alternative assessment of animals.

Animal Lovers

In his 1933 study *The Happy Beast in French Thought of the Seventeenth Century* George Boas labeled the group of writers most concerned with the idea of the natural virtuousness of animals "Theriophiles": animal lovers. For Boas, the main contributors to this tradition were Michel de Montaigne and Pierre Charron, both of whose work was available in French and English by the early seventeenth century. These writers, Boas argued, "found the true models [of men] in the animals." That is, as he puts it, "The theoretical—if not psychological—basis of Theriophily is that the beasts—like savages—are more 'natural' than man, and *hence* man's superior."[16] Put simply, Boas reads these Theriophiles as followers of Plutarch.

In the longest of his *Essais*, the "Apology for Raymond Sebond," for example, Montaigne follows Plutarch and asks why men "attribute to some sort of natural and servile inclination these works [of animals] which surpass all that we can do by nature and by art?" Animals, in his view, are neither slaves, nor inferior; in fact, men, he argues, are "without realizing it, grant[ing animals] a very great advantage over us, by making Nature, with maternal tenderness, accompany them and guide them as by the hand in all the actions and comforts of their life." Humans, on the other hand, are obliged to "seek by art the things necessary for our preservation." And art is, within this Plutarchian discourse, an addition to nature which is inferior to it. As if the human reliance on art were not bad enough, Montaigne argues, even with education humans can never reach the level of the beasts, because animals' achievements are natural and not learned, "so that their brutish stupidity surpasses in all conveniences all that our divine intelligence can do."[17] The

---

[15] Pierre Viret, *The Schoole of Beastes, Intituled, the good Housholder, or the Oeconomickes*, trans. I. B. (London: Robert Wal-de-graue, 1585), sig. 2ʳ and n.p. The text was originally published in French in 1561.

[16] George Boas, *The Happy Beast in French Thought of the Seventeenth Century* (1933; repr., New York: Octagon Books, 1966), 1.

[17] Michel de Montaigne, "Apology for Raymond Sebond," translated by Donald Frame, in *The Complete Works* (London: Everyman, 2003), 404.

terms designating animal and human—brutish, divine—may come from the discourse of reason, but the meaning certainly does not.

Going even further into the heartland of the discourse of reason, Montaigne's friend Pierre Charron asserts the authentic nature of animal's reason and goes on to criticize those who "maliciously attribute" the actions of animals that seem reasonable "to a naturall, seruile and forced inclination; as if beasts did performe their actions by a naturall necessitie, like things inanimate, as the stone falleth downward, the fire mounteth vpward."[18] Once again, free will enters the debate (if not by name, then by implication) and is opposed to instinct; for Charron, however, animals possess the former and not the latter, even though, as he notes, this is regularly missed by commentators. From two sides, then—by affirming animals' possession of both reason and free will—Charron questions the construction of animals in the discourse of reason.

This challenge to the discourse of reason is made clearest of all in Montaigne's famous question: "When I play with my cat, who knows if I am not a pastime to her more than she is to me?"[19] Who, he asks, is the object of the game here, and who the subject? What Montaigne has done—as he also did in "Of cruelty"—is turn away from the ideal animals of philosophical speculation and instead look at real beings. This cat, he insists, is his cat, not a fictional one. It is an animal in the world, and not a beast in a book. On this basis it is unsurprising that such an animal is within humanity's ethical framework.

Such a sense of the reality of the animals he is dealing with pervades Montaigne's "Apology." For example, he dismisses the suggestion (that he takes from Ovid) that humans alone look up to the heavens, with the simple anatomical statement that "there are many little creatures whose sight is wholly overturned to the sky; and I find that camels and ostriches have their necks set even higher and more erect than ours." But he goes beyond this wonderfully reductive empiricism when he refers to the guide dogs he has seen being used by the blind, and asks "How can this dog have been made to understand that it was his responsibility to consider solely the safety of his master and to despise his own comforts in order to serve him?" Montaigne's answer is in the form of another question, but this question is itself an answer: "Can all this be understood without reasoning and intelligence?" Montaigne also relates a story of the Fox of Thrace, who would put his ear to the ice on the frozen river to see if it was thick enough to walk across. But Montaigne turns this story—handed down as it is from Plutarch—into an empirical challenge. He asks, "If we saw him at the edge of the water . . . would we not have reason to suppose that there passes through his head the same

[18] Pierre Charron, *Of Wisdome* (London, 1607), 105 and 107.
[19] Montaigne, "Apology for Raymond Sebond," 401.

reasoning that would pass through ours? . . . For to attribute this simply to a keenness of the sense of hearing, without reasoning or inference, is a chimera, and cannot enter our imagination."[20] Just as he had looked at the tears of the stag and from that real relationship offered a new ethical framework for animals to his readers, so once again it is by turning to the real world of animals that Montaigne can challenge the discourse of reason.

Charron too follows Plutarch in mentioning "certaine aduantages that beasts haue ouer men," and offers an utterly conventional list that includes "Their sufficiencie in some arts, as the swallow and other birds in building, the Spider in spinning and weauing, diuers beasts in Physicke, and the Nightingale in Musicke." He too repeats the tale of the fox and the ice, and invokes the image of the dog who, chasing a hare to a crossroads, sniffed the first and the second paths, and then ran down the third without sniffing it, having reasoned, if not that, or that, then this.[21] This animal tale, attributed to the Stoic Chrysippus (c. 280–c. 206 BCE), was, as Luciano Floridi has noted, "part of the alleged evidence discussed by several philosophers who endeavoured to understand whether animals have their share of intelligence, and if so, what this may involve."[22] Another philosopher besides Plutarch who mentions the story is Sextus Empiricus, whom I return to later in this chapter.

Charron concludes by echoing Montaigne's statement that "there is more difference between a given man and a given man than between a given animal and a given man," and writes:

> we must confesse that beasts doe reason, haue the vse of discourse and iudgement, but more weakly and imperfectly than man; they are inferiour vnto man in this, not because they haue not part therein at all; they are inferiour vnto men, as amongst men some are inferiour vnto others; and euen so amongst beasts there is such a difference: but yet there is a greater difference betweene men; for (as shall be said hereafter) there is a greater distance betweene a man and a man, than a man and a beast.[23]

Charron maintains the superiority of humans but cannot offer the absolute distinction between reason and unreason. Instead, he proposes that animals reason, but less well. However, animals' inferiority to humans is nothing

---

[20] Ibid., 432, 412–13, 409. On the fox of Thrace see Plutarch, *Whether Creatures Be More Wise*, 962.

[21] Charron, *Of Wisdome*, 104, 105, 106.

[22] Luciano Floridi, "Scepticism and Animal Rationality: the Fortune of Chrysippus' Dog in the History of Western Thought," *Archiv Fur Geschichte der Philosophie* 79, no. 1 (1997): 35–36.

[23] Montaigne, "Apology for Raymond Sebond," 415; Charron, *Of Wisdome*, 108.

compared to some humans' inferiority to other humans; and this serves to upset the notion of animal inferiority. If some humans are more inferior to some humans than some animals are to some humans, does that not mean that some humans are inferior to some animals?

While Boas and others since him have concentrated on Montaigne and Charron as the key Theriophiles, and have therefore regarded this as a primarily French tradition,[24] it is possible to find Plutarchianism—and its attendant empiricism—in the work of English writers in the late-sixteenth and early-seventeenth centuries. Where many of these writers differ from Montaigne and Charron is in the inclusion of Plutarchian ideas within texts the overall desire of which is to maintain the Aristotelianism of the discourse of reason. For this reason, the texts referred to here are less thoroughgoing in their Plutarchianism than Montaigne's or Charron's work, and they are, as such, perhaps more illustrative of the struggle that was taking place in English ideas in this period.

Natural Reason

Reason, as we have seen, goes hand in hand with free will—possess one and, the discourse of reason argues, you must possess the other. By implication, those who lack reason also lack free will: as Nemesius, a sixth-century Aristotelian, wrote, "The workes of *chance* are such as befall *unreasonable*, and *inanimate* creatures, without *nature*, or *art*."[25] That is, animals, plants, and minerals are subject to fortune, whereas humans are subject only to themselves and their own deliberation. In his *Anatomie of the minde* (1576) Thomas Rogers seems to speak for this orthodox view of the incapacity of animals when he discusses prudence. For Rogers, prudence "consisteth especially in foreseeing thinges, or in preuenting a mischiefe before it come." "And therein," he continues, adding what seems to be the final statement on the matter, "doo we differ from brute, and vnreasonable creatures, which haue no forecaste, but serue the time present." This is completely orthodox. However, Rogers's orthodoxy immediately comes to invite scrutiny. In the sentence following the claim that animals only "serue the time present," he writes: "And yet we read that there is a certaine Prouidence in some beasts, as in Myce, and Antes. It is reported that by nature this prouidence is geuen to Myce, that before any man, they will foresee the destruction of an olde

---

[24] See, for example, Peter Harrison, "The Virtues of Animals in Seventeenth-Century Thought," *Journal of the History of Ideas* 59, no. 3 (1998): 470–79. Harrison does cite one English writer, John Bulwer, as a supporter of Montaigne's ideas (476).

[25] Nemesius, *The Nature of Man* (London: Henry Taunton, 1636), 542.

house, and therefore before the ruine and fall thereof, wyll leaue the same, and seeke a newe habitation." The natural world of animals, even here as it is represented by report rather than observation, enters to undermine the certainty of human difference. Rogers's response to this danger to the discourse of reason is simple: "But let vs leaue the examples of beasts, and come to men againe."[26] Having noted a contradiction, he goes on to ignore it. This is something that is not particular to Rogers.

A similar shift had already occurred in John Maplet's *Greene Forest, or a naturall Historie* (1567). Here Maplet looks at what he terms the "vertues" of the natural world. Using Aristotle as his initial guide, Maplet begins with minerals, moves to plants and then, in the third book, turns to animals, which, he argues, are different from plants in that, while both groups possess life (a crucial point for Aristotle), animals possess the abilities of "mouing from place to place, with an appetite to repast themselues." To these basic functions Maplet then adds more complex capacities: animals, he writes, have "feeling and sense"; they can perceive "griefe and paine, good and euill." This second range of qualities, Maplet states, are a "Stayer higher . . . [and] layeth holde and apprehendeth another kind of life in degree more Princely, and in force or large power most manifolde: for with this, it hath might to moue, to haue lust or appetite: to haue and hunt after what it will, and to wander and stray therefore whether it will." Animals are princely, not fully regal; that status is maintained for men who, according to Maplet, "standeth on a step higher than we [in this book] meane to climbe."[27] In this, Maplet follows convention: animals have only the organic soul, but through this they are able to perceive and (in a limited organic way) judge the world around them. Only humans go beyond the body.

Having established that there is such a distinction between animals and humans, however, Maplet then proceeds, as Rogers was to do nine years later, to draw up a list of animal capacities that would seem to undermine the difference that he has set up. Animals, in his list, are not without something akin to reason, it would seem.

> Further, some [animals] be of good memorie, or retaine for a time in their head a good turne done to them or an euill as ŷ Dog, the Lion, the Cammel: Other as forgetfull of such kinde of deseruing, and unmindeful as the Ostrich, the Doue. And in some there is a certaine kinde of perceiueraunce and adiudging or esteeming what is what (but the same spoken of vs by resem-

---

[26] T. R. [Thomas Rogers], *A philosophicall discourse, Entituled, The Anatomie of the minde* (London: Andrew Maunsell, 1576), sig. 93[r].

[27] John Maplet, *A Greene Forest, or a naturall Historie* (London: Henry Denham, 1567), sigs. 66[v] and 67[v].

blance and vnproperly) which we may perceiue they haue through their care in bringing vp & tendring their yong: as also in artificiall maner of building their Nest, in hunting and seeking after their pray, in remedying and curing woundes.[28]

These are the natural virtues of animals: bearing a (rightful) grudge, judging, nest building, rearing young. The difference between the animal and the human ability to judge is made parenthetically: animals resemble humans in their judgments, but our representation of—the language we use to describe—that resemblance is done "vnproperly." Their judgment is beyond our (human) ability to represent it.

However, the fact that, as we have seen, the inorganic soul needs training into action means that animals may compensate for a lack in one area of the debate with an advantage in another. No one, after all, teaches an animal to rear its young, whereas the number of domestic manuals and exhortations to parents made in this period suggests that humans needed constantly to be reminded of their parental duties. This is something that Thomas Wright also makes clear: all children, he argues, are born with a natural ignorance. He continues in what should by now be familiar terms:

> I must confesse one poynt of my ignorance, that it seemeth to me, that God endoweth bruite beastes with more sparkes of knowledge then reasonable men, and they may be sayde, euen in their natiuitie, to haue imprinted a certayne knowledge and naturall instinct, to inquire and finde out things necessarie, to be their owne Physitians, to flie that may hurt them, and followe that may profite them. Marke but a Lambe almost new yeaned, how it will finde foorth the mothers dugge, discerne and single her foorth in all the flocke, waite vppon her so dilligently, within eight dayes it seeth light: but a childe may be many daies borne, and yet cannot finde out his mothers dugge except the nurse moue him vnto it: neither can it cure it selfe or demaund what it needeth, other waies then by weeping.[29]

Wright's overall argument about the human passions can be situated firmly within the discourse of reason, and yet here he steps outside of that discourse and notices something that does not quite fit: how can the lamb's ability to know its mother and the human infant's failure to do the same sit comfortably alongside the assertion of human possession and animal lack of reason? Like Rogers, Wright does not dwell on this point, but in noting it he undoes the simple assertion of the inferiority of animals that lies at the

[28] Ibid., sig. 68ʳ.
[29] Th. W. [Thomas Wright], *The Passions of the Minde* (London: V. S. for W. B., 1601), 228.

heart of his text. For all of these thinkers, it seems, animals have skills naturally whereas humans have them only culturally, by habit, and this is a feature of the natural order that cannot be easily avoided or easily accounted for by the discourse of reason.

Perhaps the most extended illustration of the influence of Plutarch in early modern England comes in the work of Godfrey Goodman, the Bishop of Gloucester. Goodman's *Fall of Man* (1616) begins with some fairly orthodox statements: "Man, without education," he writes, "is like the dumbe beast, sauage and wilde." But he is also "alone of all other creatures, in regard of the freedome of his will, and the choyce of his owne actions." Animals, by implication, are incapable of choice, and therefore subject to chance. However, just as Rogers, Maplet, and Wright had done before him, so Goodman highlights a limitation to human superiority. Goodman differs from the earlier writers, however, in that he does not evade the issue once he has named it; he examines it.

Goodman's link to Plutarch is made clear in his assessment of animal capacity:

> Other creatures excell man in euery sense, in euery corporeal qualitie, as length of yeeres, strength, soundnes of constitution, quicknes, actiuitie; man cannot be so subtill and ingenuous to insnare thē, but they are as cunning and wittie to preuent vs; you may assoone surprise and conquer a State, as preuaile against them in this kinde.

For Plutarch, the difficulty of catching fish was evidence of their reason. Goodman makes another distinction between humans and animals that also echoes Plutarch:

> I speake not of their aptnesse in learning, which I haue seene wonderfull in horses, and in other creatures, for this I might ascribe in some sort to their teacher: but I speake of their naturall workes; the birds in building their nests, with straw in their bils, water in their wings & earth in their clawes, may serue to teach and instruct vs in our architecture, in the plot-forme and contriuing of our houses.

Art does not set human apart from animal—does not reveal humanity's capacity to reason—rather, human art is figured as a poor imitation of animal nature. Goodman writes, "The little chirping birds (the Wren, and the Robin) they sing a meane; the Goldfinch, the Nightingall, they ioyne in the treble; the Blacke bird, the Thrush, they beare the tenour; while the foure footed beasts with their bleating and bellowing they sing a base." This natural music of animals is set against the unnatural attempts of the human, whom Goodman presents as "a wild and a fierce creature." Human music, he writes, "hath no certaine note or tune . . . his instruments are the guts of dead creatures, a token of his crueltie, and the remainder of his riot." Goodman then goes on to ask a question, the source of which must be Montaigne.

> For the pleasures and sports belonging to bruite beasts, you see that Princes and Nobles take their greatest pastime, in those royall games of Hawking, and Hunting. I would gladly know, whether the Faulcon receiues more delight in the sport, or the Faulconer?

For Goodman, as for Montaigne, the animal is not an object for use but a subject in its own right. It is capable of experiencing pleasure, like the human. And more than this, the hawk's pleasure in hunting is more natural than the human's because the hawk is "an actor in the businesse, it being more agreeable to the nature thereof; the other a bare spectator in the game."[30] In Goodman's vision of hawking, in fact, the hawk acts while the humans suffer. This is very different from the representation of animals in the discourse of reason.

Whereas John Donne, following the logic of the discourse of reason, called for the disafforestation of the human mind, Plutarch and his early modern followers seem to desire that the wilderness be reinstated. But their wilderness is not a place of violence and struggle; it is a place of a natural and reasonable capacity, unfettered by civilization and education, where music is made by nature and is not taught or played on the bodies of dead animals. This overturning of the orthodox vision of animals is formulated in an attempt to truly comprehend what it is that animals do—when hounds follow the scent, when guide dogs display selfless comprehension of the needs of their owners. Such interest opens up the possibility of a new way of thinking about their organic capacities. The existence of a classical source that supports animal rationality perhaps gives early modern writers more confidence in their reassessment of animals' actions, but it also allows what it is that people are actually seeing to enter into the domain of theoretical speculation. This is made clear in a debate that took place at Cambridge University in 1615, where a highly placed participant appeared to claim more for animals than the discourse of reason allowed.

More Than Is Imagined

The subject of the debate that took place at Cambridge University in 1615 was "tempered & fitted" to suit the king who would be in attendance; and

---

[30] Godfrey Goodman, *The Fall of Man, or The Corrvption of Nature, Proved by the light of our naturall Reason* (London: Felix Kyngston, 1616), 25, 27, 76, 78, 87.

James VI and I's love of hunting led to the choice of a particular issue, "namely whether Dogs could make syllogismes"; that is, whether dogs could engage in complex reasoning. The puritan John Preston was engaged to propose the affirmative, and he cited Chrysippus when he "instanced in a Hound, who hath ye major proposition in his minde, namely, the hare is gone either this way, or that way, smells out the minor wth his nose, namely, she is not gone that way, & follows the conclusion, 'Ergo,' this way, wth mouth open."[31] In his record of the debate Thomas Ball wrote that the image of Chrysippus's dog "suited wth ye Auditory and was applauded." The Answerer in the debate, Matthew Wren, the future Bishop of Ely, was "put . . . to his distinctions," and he proposed, in orthodox fashion, "that dogs might have sagacity, but not sapience, in thinges especially of Prey, and that did not concerne their belly." Wren made a distinction between the capacities "nasutuli" and "Logici"—that is, between the ability to smell and the capacity to think—arguing that dogs "had much in their mouths, little in their myndes, unless it had relation to their mouths, that their lips were larger than their understandings." This, of course, is utterly conventional.

Preston's response was, however, cut short by the moderator of the debate, Dr Reade, who instead removed the hearers into the realm of metaphor. Reade

> began to be afraid, and to think how troublesome a pack of hounds well followed and applauded at last might prove; and so came in unto the Answerer's [that is, Wren's] Ayd, and told the Opponent that his dogs he beleeved were very weary, and desired him to take them off; and when the Opponent would not yield, but hallowed still and put them on, he interposed his authority & silenced him.

The debate, it appears, was not taken wholly seriously; it was merely an exercise in philosophical disputation, and the issue of animal intelligence was not deemed worthy on this occasion of significant discussion.

But the debate in Cambridge does not end here. There is another intervention, and a crucial one.

> The King, in his conceit, was all this while upon New Market Heath, & liked the sport; and, therefore, stands up and tells the Moderator plainly he was not satisfied in all that had bin answered, but did beleeve a hound had more in him than was imagined. I had myself (said he) a dog that stragling farr

---

[31] Thomas Ball, *The Life of the Renowned Doctor Preston, writ by his pupil, Master Thomas Ball, D. D. Minister of Northampton, In the Year 1628,* ed. E. W. Harcourt (Oxford and London: Parker & Co., 1885), 20, 21, and 23.

from all his fellows had light upon a very fresh scent, but considering he was all alone and had none to second and assist him in it, observes the place & goes away to his fellows, and by such yelling arguments as they best understand, prevayled wth a pty of them to goe along wth him, and, bringing them to the place, psued it into an open view. Now the King desired for to know how this could be contrived and carried on without an exercise of understanding, or what the Moderatour could have done in that case better, & desired him that either he would thinke better of his dogs or not so highly of himselfe![32]

James has shifted the focus of the debate from philosophical speculations about canine syllogizing into the realm of empirical observation: he refuses to allow the discussion to be positioned as comic or merely theoretical, but instead recognizes that it could be a real issue if applied to real dogs. Just as real animals begin to undo the construction of the human in the discourse of reason when the beastliness of humanity is brought to the fore, so here emphasis on real animals offers the opportunity to ask another potentially destructive question: When do an animal's actions stop being merely sensual and become something more like the reason that is expressed by humans? When, to use Chrysippus's image, does a dog following a hare cease to use mere instinct and begin to syllogize? Ultimately, for James VI and I, the answer is that the two—instinct and syllogism—are not unrelated; *instinctive* reasoning might be something that animals can undertake.

The shift from the ideal to the real—the merely speculative to the empirically observed—made by King James is, however, politely silenced, despite John Preston's desire to "psue the King's game." Thomas Ball writes, "But the Answerer [Wren] protested that His Majesties dogs were always to be excepted, who hunted not by cōmon law but by prerogative." And the moderator, "fearing the King might let loose another of his hounds, and make more worke, applyed himself wth all submissive devotion to the King, acknowledged his dogs were able to outdoe him, besought His Majesty to beleeve he had ye better." It is by veering away from discussions of animals and into discussions of royal influence and prerogative that the debate is brought to a close. Revealing himself, perhaps, to be more interested in his own power than in the power of his hounds, James "went off well pleased wth ye business."[33]

This move from real animals to symbolic ones made by the answerer and the moderator of the debate is echoed twenty years later in *An Occasionall Discourse, upon an Accident which befell his MAIESTY in Hunting*. In this brief

[32] Ibid., 23–24, 25–26.
[33] Ibid., 25–26.

text Francis Cevolus tells the story of an event of 1633, in which Charles I, "being attended with a glorious traine of Noble Courtiers, a pursued Hare runne for protection to his sacred Person, and freed herselfe from an eminent danger by taking sanctuary under his Horses belly." This "nine dayes wonder" made a great "impression" on Cevolus, and led him to contemplate whether "that little creature might afford some considerations, not misbeseeming so great a Majesty." The considerations that followed were not about the hare so much as about the king:

> The Hare in her kind is endowed with a quicke hearing: which may well betoken the Kings facility in affording audience to the meanest suiters. The Hare is observed to sleepe with her eyes open. . . . The King never slumbers in the affaires of his people, but watcheth over them that they may sleepe secure; and buyes their rest by his owne disquiet.[34]

We have here another appearance and disappearance of real animals. In previous chapters real animals disappeared—to be replaced by symbols—as their presence became too great a challenge to the superiority of humans. In the Cambridge University debate, and in Cevolus's occasional text, real animals are initially worthy of contemplation, but only insofar as they draw the human from the material world and into contemplations and abstractions. On this basis their status as beings worthy of study in and of themselves is, to say the least, limited. In fact, to stick with the real animals would challenge the distinction of humans, since the actions of those real animals appear to be somewhat reasonable.

As can be seen in the shifts in focus outlined so far in this chapter, in early modern England there was no one way of viewing animals. Using Plutarch and Montaigne as guides, thinkers could propose the superiority of animals; by studying animal behavior, they could offer challenges to the assertions of the discourse of reason. However, persisting alongside this—and visible in both the conclusion of the 1615 debate and the story of Charles I and the hare—is the discourse of reason; a mode of representing animals that supports human superiority. In fact, the discourse of reason is maintained in discussions of animals by, paradoxically, focusing those discussions on humans. This can be seen in what appear to be two very distinct discourses; namely, in meditations and in natural philosophy.

---

[34] Francis Cevolus, *An Occasionall Discourse, upon an Accident which befell his MAIESTY in hunting* (London: Iohn Norton, 1635), 3, 5.

Ideal Animals

In *The Arte of Divine Meditation* (1607) Joseph Hall outlines an orthodox vision of animals.

> Man is placed in this Stage of the world to viewe the seuerall natures and actions of the creatures; To view them, not idly, without his vse, as they doe him: God made all these for man, and man for his owne sake; Both these purposes were lost, if man should let the creatures passe carelesly by him, onely seene, not thought vpon: He onely can make benefit of what he sees; which if hee doe not, it is all one, as if hee were blinde, or brute. Whēce it is, that wise *Salomon* puttes the sluggard to schoole vnto the Ant; and our Sauiour sends the distrustfull to the Lillie of the field.

As is clear in the last sentence, the term "creatures" refers to any and all aspects of created nature. But this general meaning of "creatures" does not detract from the idea that animals in particular have a role to play in human life that goes beyond their use as food, as workers, and so on. Not only are animals made for humans, but those animals also bear a meaning the reading of which helps to construct the distinction between human and animal. Hall goes on, "The brute creatures see the same things, with as cleare perhaps better eyes: if thine inward eyes see not their vse, aswell as thy bodily eyes their shape, I knowe not whether is more reasonable, or lesse brutish."[35] Here, mere observation encounters the emphasis made by the discourse of reason on the capacity of humans to think in abstract terms, and what emerges are animals meaningful with respect to issues that have only a tangential relationship to those animals themselves. Animals, in fact, are prompts for wider discussions that exclude animals and define humans.

Hall's outline of the role of animals (and other "creatures") in meditation is repeated in numerous texts from the period. Hall himself reflects on a robin, a spider, a barking dog, snails, flies gathering on a "galled Horse," and a dormouse in his *Occasionall Meditations* (1630).[36] Likewise, writing in 1633, Edward May contemplated a toad, a dwarf on the back of a mastiff, and a ram, and the following year Donald Lupton meditated on a toad and a dog on a chain.[37] In all these cases the meditation begins with the animal

[35] Joseph Hall, *The Arte of Divine Meditation: Profitable for all Christians to know and practice* (London: H. L., 1607), 11–12, 22–23.

[36] Ios. Exon. [Joseph Hall], *Occasionall Meditations. Set forth by R[obert] Hall* (London: Nath. Butter, 1630), 32–34, 35–37, 57–59, 74–75, 78–79, 146–48.

[37] Edward May, *Epigrams Divine and Morall* (London: I. B., 1633), sigs. B4ᵛ, B7ʳ, and D6ʳ; Donald Lupton, *Obiectorvm Redvctio: Or, Daily Imployment for the Soule* (London: John Norton, 1634), 6–7, 24–26.

and then turns to a wider spiritual truth that the material being sets off in the mind of the meditator. Thus in "*Of a* Dogge *in a* Chaine" Lupton moves from an encounter with a chained and barking dog—"though being *chain'd,* he cannot *bite* with his *teeth,* yet his *barking* showes what he would doe at *liberty.*" This leads him to a contemplation of the "great *Dogge* of *Hell,*" which, "when he persecutes by *bonds, imprisonment,* and *captivity* then he *bites* sore: when he *slanders, reviles,* and *envies,* then he *snarles,* and *barkes* onely." Lupton continues and extends the analogy between the dog and Satan, and writes, "If God should not permit *this* Curre, few would feare his *justice:* if *he* should not *limit* him, many would *question* his *mercy.*" He concludes the meditation by acknowledging the power of God and the frailty of humans: "My prayer to God shall be, to tie him up *shorter,* and I could wish he were alwaies *musled,* but Gods *will* be *done.*"[38] The dog as dog has disappeared from the meditation, and what emerges in its place is a spiritual truth and, from that abstraction, the human status of the observer. This mode of perception is not only to be traced in meditations or other occasional writings, however; it is even found in natural philosophy, a discipline that supposedly had animals as its main focus.

In his *Historie of Foure-Footed Beastes* (1607), for example, Edward Topsell, while ostensibly cataloging the names, anatomies, and myths surrounding many of the quadrupeds with whom humans shared their world, had a clear intention: he proposed that animals were created in order "that a man might gaine out of them much deuine knowledge, such as is imprinted in them by nature, as a tipe or spark of that great wisedome whereby things were created."[39] Similarly, in the unfinished third part of his survey of animals, *The Fowles of Heauen* (c. 1613–14; his 1608 study of "serpents" was the second), Topsell wrote that "the scriptures cannot be rightlie vnderstood without the knowledge, and historie of fowles."[40] For Joseph Hall, animals are like the "capitall letters of Gods great booke," and it is the duty of humans to learn to "spell"; to make sense with these "letters" that God has provided.[41] Topsell likewise writes,

> When I affirm that the knowledge of Beasts is Deuine, I do meane no other thing then the right and perfect description of their names, figures, and na-

[38] Lupton, *Obiectorum Redvctio,* 24–26.

[39] Edward Topsell, *The Historie of Fovre-Footed Beastes* (London: William Jaggard, 1607), sig. A4ʳ.

[40] Edward Topsell, *The Fowles of Heauen or History of Birdes* (c. 1613–14), ed. Thomas P. Harrison and F. F. David Hoeniger (Austin: University of Texas Press, 1972), 9. The second part of his trilogy of texts is Topsell, *The Historie of Serpents. Or, The second Booke of liuing Creatures* (London: William Jaggard, 1608).

[41] Hall, *Arte of Divine Meditation,* 22.

tures, and this is in the creator himself most Deuine, & therfore such as is the fountain, such are the streams yssuing frō the same into the minds of men.[42]

Thus, in terms of the understanding of animals the distance between meditation and natural philosophy—between Hall and Topsell—is not great. To study the natural world, in fact, is to study not animals, plants, or minerals but God.

Natural philosophy also reproduces meditation's emphasis on the centrality of the human. In line with the notion of the chain of being, animals were understood to exist beneath humans and for humans, and this perception clearly included any outlining of their behavior. As Peter Harrison has written, in early modern natural philosophy "the literary context of the living creature was more important than its physical environment. Animals had a 'story,' they were allocated meanings, they were emblems of important moral and theological truths." In short, animals were prompts to the abstract.

As well as symbolizing moral truths, however, animals were also used to concretize that fragile being called the human: Harrison notes that the human was perceived as "an epitome of all the animals. Birds and beasts could thus symbolise distinct passions, virtues and vices."[43] The cunning of a fox, the loyalty of a dog, the timidity of a hare, all of these apparently *predetermined* (that is, not learned) animal behaviors were used to explain more generally the concepts of cunning, loyalty, and timidity in humans. The fox, the dog, and the hare were, in themselves, of only marginal interest because their activities were somehow not theirs at all but belonged to nature and to God: animals suffer, they do not act.

In these terms, an animal was represented as meaningful and recognizable to humans more as an exemplar of a particular human trait than as a fellow being—Peter Harrison has argued that animals were "cyphers, insignificant in themselves, yet useful for humans at every level."[44] To offer just one example of this trope of natural philosophy, Topsell, using Pliny's *Naturall Historie* as his key source,[45] begins his chapter "Of the Elephant" with the following statement: "There is no creature among al the Beasts of the world which hath so great and ample demonstration of the power and

[42] Topsell, *Historie of Foure-Footed Beastes,* sig. A4r.

[43] Peter Harrison, *The Bible, Protestantism, and the Rise of Natural Science* (Cambridge: Cambridge University Press, 1998), 2, 185.

[44] Harrison, "The Virtues of Animals," 468.

[45] An English translation was of this text was published in 1601: *The Historie of the World. Commonly called, The Naturall Historie of C. Plinius Secundvs,* trans. Philemon Holland (London: A. Islip, 1601). The discussion of elephants forms the first twelve chapters of Book VIII.

wisedome of almighty God as the Elephant: both for proportion of body and disposition of spirit." The spirit of this animal includes its generosity: "They are so louing to their fellowes, that they will not eat their meat alone, but hauing found a prey, they go and inuite the residue to their feastes and cheere, more like to reasonable ciuill men, then vnreasonable brute beasts."[46] Here the elephant (figured, it seems, as a carnivore) offers to Topsell's readers a vision of how a good human might behave.

Gail Kern Paster has recently offered a different interpretation of the role that animals play in early modern analysis that runs counter to mine. Because humans and animals share in possession of the passions, housed as they are in the body rather than the mind, early modern discussions of animals, Paster argues, should not be interpreted as anthropomorphic, but as recognizing this shared aspect of existence. For example, she proposes that, in Topsell's *Historie of Foure-Footed Beastes,*

> there is not only significant continuity between human and animal emotions but also . . . a descriptive vocabulary in which ethical, physical, and psychophysiological discourses intermix. . . . For Topsell, as for other early modern thinkers, it was not just that the qualities of animals resembled those of human beings, but that those qualities were directly transferable from animal to human as humans applied and incorporated animal flesh into their own.

In Paster's analysis, to label such writings anthropomorphic is to miss the fact that when, for example, the cat is regarded as "melancholy," it is because a cat is perceived to share the corporeality that can produce melancholy; that it is not a false transfer of a human trait onto a creature incapable of such a humor.[47] Paster is certainly correct to note the shared nature of such corporeal states as the humors and the passions, but she does not consider that Topsell and other early modern writers are often very far from being interested in the animals themselves. God has sent animals as a sign, and the natural philosopher's argument is twofold: first, he believes that it takes a religious man to interpret the natural world correctly (Gordon L. Miller has written of Topsell's *Fowles of Heauen,* "it took a good heart to know the truth about cranes"[48]); and second, that humans should learn to interpret the nat-

---

[46] Topsell, *Historie of Fovre-Footed Beastes,* 190 and 196.

[47] Gail Kern Paster, *Humoring the Body: Emotions and the Shakespearean Stage* (Chicago: University of Chicago Press, 2004), 154, 135–88 passim.

[48] Gordon L. Miller, "The Fowles of Heaven and the Fate of the Earth: Assessing the Early Modern Revolution in Natural History," *Worldviews: Environment, Culture, Religion* 9, no. 1 (2005): 68.

ural world correctly and from that interpretation become better—more Godly—people. Once again, knowing nature is, importantly, about knowing oneself.

The outcome of this understanding of the study of animals is, then, that animals are perceived through the filter of a human vision and for a partic ular (human) purpose. And while, as Paster has pointed out, to simply label the forms of representation used in natural philosophy anthropomorphic might be to overlook the important linking of human and animal that lies at the heart of faculty psychology, it is hard not to regard the "ciuill" elephant, for example, as evidence of anthropomorphism. Although this anthropomorphization of animals might appear to reduce the distance between humans and animals, it does not uproot difference; in fact, it is the existence of the anthropomorphic nature of animals that—paradoxically, it might seem—stabilizes the natural hierarchy, since anthropomorphism reiterates the centrality of humans in the natural order. Tom Tyler has written, "To be able to claim that a characterization or representation of some being assigns to it a quality or state that is distinctively human, one would need to know just what it is about human beings, in themselves that makes them the kind of beings they are."[49] By anthropomorphizing animals—by regarding the elephant as a creature capable of "ciuill," and thus reasonable, behavior, for example—what emerges is not simply a picture of an animal perhaps closer to the human than might have been expected, but a fixed and stable concept of the human, from which all other "creatures" in the world can be interpreted.

And, by implication, within this framework there is little harelike about a hare that is worth noticing unless that harelikeness enhances the observer's understanding of the human. Thus Topsell's work operates in an orthodox way within the discourse of reason in that his animals are the means of reiterating human nature and centrality. A "ciuill" elephant within this discourse is a useful elephant, and while "use" might usually apply to an animal's edibility, trainability, and so on, it can also be applied to humans' ability to recognize and understand that animal because by doing that humans are, in fact, recognizing and understanding their (ideal) selves.

Factual Fictions

By using animals to fully understand themselves, then, humans constructed animals as meaningful. The meditator shifted from a chained dog to the per-

---

[49] Tom Tyler, "If Horses Had Hands . . . ," *Society and Animals* 11, no. 3 (2003): 273.

secutions of the devil, and the natural philosopher traced a short step between animal actions and the stability of the concept "human," and so it becomes clear that the categories used to define and explain animals are categories that center on the human. This is something that can also be traced in a work such as William Harrison's *Historicall Description of the Islande of Britayne* (1577), which outlines the "sauage beastes and vermines" of the island.

Harrison begins his discussion of animals by noting the blessing bestowed on Britain in its lack of "Lions: Beares, Tygers, Pardes, Wolfes." He then outlines what verminous animals do exist there and includes under this heading foxes, badgers, polecats, stoats, weasels and squirrels, and notes that there are very few beavers in Britain—"only in the Teifie in Wales"—whereas otters are to be found "in many streames & riuers." This survey does not include much detail of the animals' actual behavior—how they live—rather, the focus is on the basic fact that these animals do live. Their mere presence is all that is recorded, because, it could be argued, animals are only ever merely present; they are never, in this discourse, actively present.

Such brief descriptions of animals occasionally slide into something rather different, however. Harrison's depiction of the stag, the fallow deer and the hare, for example, leads into the following:

> Of these also the stagge is accompted for the most noble game, the fallow Deare is the next, then the Roe (whereof wee haue indifferent store) and last of all the Hare: all which (notwythstanding our custome) are pastimes more meete for Ladies and Gentlemen to exercise, then for men of courage to followe, whose hunting should practise theyr armes in tasting of theyr manhoode and dealing w such beastes as eftsoones wyll turne agayne and offer them the hardest, rather then theyr feete, whych many tymes may cary dyuers from the fielde.

From an evaluation of animals to a debate about the uses of hunting, Harrison's discussion is only partially interested in the animals as animals. Rather, human benefit is central. When he arrives at his survey of "Cattell kept for profite," however, even this partial interest has vanished. This is the entire entry on English pigs:

> As for Swine there is no place that hath greater store nor more wholsome in eating, thē are here in England & of these, some we eate greene for porcke, & other dryed vp into Bacon to haue of it more continuance, Larde we make little because it is chargeable, neyther haue we such vse thereof as is to be seene in other Countries, sith we do either bast all our meate with butter, or suffer the fattest to baste it selfe by leysure.

Boars are likewise discussed only in terms of the making of brawn.[50] The living animal is completely ignored, and in its place its meat is the focus. Harrison makes no effort to signal this shift because, for him, it is not so great; animals were always objects anyway; so it is their use to humans and not their innate worth that gets recorded.

Similarly, in his 1602 *Survey of Cornwall* Richard Carew offers what appears to be a factual overview of the animals to be found in that county. He begins at the bottom of the chain of being—with worms, snakes, and rats—and goes on to look at "other beastes which *Cornwall* breedeth, [that] serue either for Venerie, or meate, or necessary vses."[51] The taxonomical distinction that Carew makes is between useless (vermin) and useful (huntable, edible, working) animals, and thus he replicates the categories that are implicit in Harrison's work. Such categories only make sense if humans are regarded as the organizing principle of the world; if animals themselves are perceived to be interpretable only through their relation to humans. In this respect, Carew's work is utterly orthodox, and there is accordingly almost no description of the activities of animals (since such information would be unimportant). These apparently real animals are, in fact, revealed to be ideal; they allow the writers to reproduce the orthodox construction of human–animal relations, in which animals themselves are only of marginal significance.

While Harrison's and Carew's surveys might offer little discussion of animal behavior, however, that does not mean that there was no interest in it in the early modern period. Indeed Jason Scott-Warren has argued that "a large part of the pleasure of blood sports for the early modern viewer had to do with what it revealed about the animals." He proposes that "the bearpits and cockpits enabled animals to become objects of knowledge, exposing their inner natures to outward view."[52] Scott-Warren is right to note that animal combats were frequently represented as having this strangely "educational" quality to them: in his *Commendation of Cockes, and Cock-fighting* (1607), for example, George Wilson argued, "let not vs which are men, indued with wisdome and vnderstanding, and with farre greater prerogatiues of nature, then any other inferiour creatures, they being subiected vnto vs, and made for our vse: but we hauing our natures much indeered, and bettered by art, Let not vs I say, shew more cowardize and faint hearted timorousnesse, then these silly fowles of the air haue done." To watch a cock-

---

[50] William Harrison, *An Historicall Description of the Islande of Britayne,* in Raphael Holinshed, *The Firste volume of the Chronicles of England, Scotlande, and Irelande* (London: Iohn Harrison, 1577), fols. 108r–109r and 110r.

[51] Richard Carew, *The Survey of Cornwall* (London: SS, 1602), sig. 22r–v.

[52] Jason Scott-Warren, "When Theaters were Bear-Gardens; or, What's at Stake in the Comedy of Humors," *Shakespeare Quarterly* 54, no. 1 (2003): 71 and 74.

fight is—ideally—to be moved to bravery; is to witness the cocks not as animals but as symbols of manly virtue.

This pedagogic value of the spectacle of animal combat is rather different from Viret's return to the "*Schoole of Beastes.*" In Wilson's work, animals are inferior in that they are emblematic: that is, one-dimensional. But such an assessment represents only half the story. The animals' value is premised not on witnessing and making sense of the behavior of animals so much as on applying preexisting traditions to that behavior. Wilson suggests that the cocks show "incredible valour . . . for no other cause (so farre as I can coniecture) then for the loue of their Hens."[53] It is clear that Wilson's "coniecture" is based on an ideal perception of manly virtue which is then transferred onto the animals. To say that cocks love their hens is to see in advance of observed fact, is to construct the animal in the absence of the animal; and so the viewer witnesses the combat not to learn about the behavior of animals so much as to learn the truth about humanity, because that is what is always already believed to be the point of the animals' behavior. Scott-Warren may be right, then, to argue that spectators watched animal combats in order to understand the animals' inner natures, but he does not fully explore the possibility that the "inner nature" of an animal was a fiction already established as fact within the discourse of reason—a fact that no amount of observation would challenge.

Such an assessment of animal combats can be traced, I think, in records of baiting matches that Scott-Warren refers to, and which are found in John Stow and Edmond Howes' *Annales, Or Generall Chronicle of England* (1615). Here Stow and Howes report on various combats attended by members of the royal family at the Tower of London, home to the king's menagerie of wild animals. In 1605, for example, they record James VI and I and some of his court visiting the Tower to watch the lions. A lion and lioness, Stow and Howes record, were "forced out" into the "walke, . . . and when they were come downe . . . they were both amazed, and stood looking about them, & gazing vp into the ayre." The lions were fed with "two rackes of mutton . . . then was there another liue Cocke, cast vnto them, which they likewise killed, but suckt not his blood."[54] This appearance of the cock seems bizarre. A cock, while an image of manly virtue in cockfighting, can hardly be said to be an ideal combatant for a lion. However, there is, I think, an explanation for this strange confrontation; an explanation that reiterates the significance of preexisting conceptions of animals for the spectators' un-

[53] George Wilson, *The Commendation of Cockes, and Cock-Fighting* (London: Henrie Tomes, 1607), sig. B2ᵛ.

[54] John Stow and Edmond Howes, *The Annales, Or Generall Chronicle of England* (London: Thomæ Adams, 1615), 865.

derstanding of what animals did. In a convention that was a part of medieval tradition, there was a belief in the natural antipathy of the lion and the cock. One twelfth-century bestiary, for example, put it simply: "A lion fears a cock, especially a white one."[55] This lore survived into the early modern period: in 1586, Geffrey Whitney wrote of "The crowinge cocke, the Lion quakes to heare."[56] And almost fifty years later the poet Richard Brathwait wrote of "The Lyon" that "He hath an Antipathy with the Cocke, especially of the Game."[57] The lions' killing of the cock at the Tower and their refusal to drink its blood could be counted as evidence to support the tale. So the lowering of the cock into the lions' den invites the royal spectators not so much to watch nature in action as to watch myth: the actions of the animals are made meaningful by their compatibility or otherwise with existing human tradition.

The challenge of the cock and the lions was not the end of the combats witnessed by the royal party in 1605, however. The spectators' investigations continued, and what occurred next once again requires some explanation. Stow and Howes write:

> After that the Kinge caused a liue Lambe to be easily let downe vnto [the lions], by a rope, & being come to the grounde, the Lambe lay vpon his knees, and both the Lyons stoode in their former places, and only beheld the Lamb but presently the Lambe rose vp, and went vnto the Lyons, who very gently looked vppon him, and smelled on him, without signe of any further hurt, then the Lambe, was very softly drawne vp againe in as good plight, as hee was let downe.

This cannot be simply understood as instigation to a sporting contest. Rather, there are, I think, three possible readings to be made. First, that the royal party merely wanted to see savagery in action, to see the lions dispatch the lamb. Second, the lowering of the lamb into the lions' den might have been an attempt by the royal party to witness a moment of leonine kindness such as Daniel experienced (Daniel 6:22) with the lamb emblematic of that other Lamb of God. Third, while this amazing spectacle may have challenged expectations about the violence of the lions, it would have, perhaps, reinforced the belief in the divine power of the overseeing monarch to control nature. Whereas Daniel cried out to "the living God," perhaps the lions' kindness in 1605 could be offered as evidence of the presence of God's rep-

---

[55] Bestiary in T. H. White, ed., *The Book of Beasts: Being a Translation from a Latin Bestiary of the Twelfth Century* (Stroud: Alan Sutton, 1992), 11.

[56] Whitney, *Choice of Emblemes*, 52.

[57] Richard Brathwait, *A Strange Metamorphosis of Man, transformed into a Wildernesse* (London: Thomas Harper, 1634), sig. B1ᵛ.

resentative on earth. Just as Charles I, twenty-eight years later, could offer protection to a hare, so James's mere presence could put an end to the violence of animals. However, this kindness of the lions was brought to an abrupt end when another lion was brought in to the "walke" and was baited by dogs. Here, perhaps, we witness the "simpler" pleasure of combat.[58] But this simpler pleasure, we should remember, is not the focus of the king's visit; it is merely its epilogue.

Thus, in attempting to read the animals through the lens of preexisting narratives—the bestiary tradition, the Bible, and the belief in the monarch's divine power—the animal combats witnessed by the king and other members of the royal family do not so much reveal the animals' natural behavior as the preconceived ideas about animals held by the spectators. The framework of myth sits comfortably beside the spectators' desire for spectacle and cruelty. Thus, what Stow and Howes record are in part human expectations and their realization (or otherwise) in the face of real animals. There are, at work in the Tower, narratives about ideal animals, and there are real animals; and the two are not always compatible.

So, when Scott-Warren argues that "The arena became a kind of psychological anatomy theater, revealing the courage, nobility, and artistry, the 'peculiar or proper' character of the animals that were exposed to the public gaze," this does not fully reflect the context in which animals were seen.[59] People looked at animals, without doubt, but they frequently did so with certain expectations of what animal nature would be. In fact, for many writers and observers those expectations constructed the meaning of the animals' actions.

This is an early modern technique of reading animal behavior that directly contradicted ideas found in another classical tradition, and it is significant for us that these ideas—or the most coherent reproduction of them—were rediscovered during the sixteenth century and first printed in 1562. Sextus Empiricus's *Outlines of Scepticism* offered early modern writers yet another way of thinking about animals. In the debates already covered in this chapter, animals were reasonable beings; were creatures capable of performing syllogisms; were beings that offered humans a way of exercising their humanity and of concretizing their centrality in the universe; and animals were organisms constructed by human stories. In skeptical analysis the nature of the debate shifted. Skepticism challenged not only the centrality of reason as a vital organizing structure of the natural order but the meaning of the human construction of animals in the discourse of reason as well. Unlike Aristotle, Sextus Empiricus was not interested in arguing for the su-

---

[58] Stow and Howes, *Annales*, 865.
[59] Scott-Warren, "When Theaters were Bear-Gardens," 74.

periority of humans; and unlike Plutarch, he was not interested in arguing for the superiority of animals. What Sextus offered to early modern writers was a reassessment of the nature of reason itself, and animals had a major role to play in this reassessment.

Skepticism

Pierre Charron's work differs from Montaigne's in one obvious way: whereas Montaigne began with the comparison of humans and animals in the "Apology for Raymond Sebond," and progressed to the difficult discussion of man himself, Charron began with a study of the *"humane condition"* in his work *Of Wisdome,* and turned to animals only in the second part of the text. As he put it, "Wee haue considered man whollie and simplie in himselfe: now let vs consider him by comparing him with other creatures, which is an excellent meanes to know him. This comparison," he goes on, "hath a large extent, and many parts that bring much knowledge of importance, and very profitable, if it be well done." But at this point in his text Charron asks a fundamental question about the comparison of humans and animals, one that had not been asked regularly in philosophical thought before him: "But who shall doe it? Shall man? He is a partie and to be suspected; and to say the truth, deales partially therein."[60] By implication, it is unsurprising that orthodox evaluations of humans and animals always assert the superiority of the human because it is, after all, always humans who are performing the evaluation. This idea of the partiality of the human vision also emerges in Montaigne's work: "How does [man] know . . . the secret internal stirrings of animals? By what comparison between them and us does he infer the stupidity that he attributes to them?" In answer to his question, Montaigne wrote—as Charron was to do—of "man's impudence with regard to the beasts," and illustrations of this impudence and evidence of animal intelligence make up the basis of much of the first half of the "Apology."[61]

George Boas interpreted the illustrations of animal intelligence in the "Apology" in terms of their relationship, primarily, with Plutarch; and while Plutarch, as I have argued above, is clearly significant, Boas fails to address the importance of skepticism to both Montaigne's and Charron's thinking about animals. The issues of human partiality and impudence that both thinkers raise threaten absolutely the ways in which writers within the discourse of reason divided the human from the animal. Montaigne and Charron, in fact, make the dividing lines that had seemed so natural in that

[60] Charron, *Of Wisdome,* 101–2.
[61] Montaigne, "Apology for Raymond Sebond," 401, 402.

discourse appear unnatural. If we read as *skeptical* Plutarchians, that thing labeled "instinct" is not the natural, truthful way of being in the world; it is an epistemological category worthy of some detailed—and destructive—scrutiny.

The urtext of early modern skepticism is, as noted, Sextus Empiricus's *Outlines of Scepticism*. Written in the latter half of the second century CE, this text was first published in Latin translation by Henri Estienne in 1562, when, Jonathan Barnes argues, "It made a sensation."[62] According to Barnes, the rediscovery of Sextus Empiricus's text was one of the reasons for the massive philosophical shift from metaphysics to epistemology in the seventeenth century. This movement from determining "what there is" to determining "what we can speak of" (to borrow Barnes's representation of the shift) places a new focus on the being doing the speaking.[63] *Nosce teipsum* (know thyself) takes on a new urgency: the question "What do I know?" Montaigne writes, is the best way of conceptualizing skepticism.[64]

Sextus argued that in philosophy there are three central schools of thought: the Dogmatists (Aristotle and his followers), who "think that they have discovered the truth"; Academics (also known as Academic skeptics), led by Cicero, who "have asserted that things cannot be apprehended"; and the skeptics (also known as Pyrrhonic skeptics), who "are still investigating."[65] Sextus's aim is to outline the last position, and he does this first of all by setting out some general principles, then turning to the logical (or illogical) constructions of the other schools.

The school of skepticism that Sextus outlines is significantly different from the academic skepticism that Cicero presented in *The Academics*. In that text Cicero stated that "we do not deny that something of the nature of truth exists, but we do deny that it can be perceived."[66] While Sextus agrees that truth cannot be perceived, he disagrees with Cicero's assertion that it still exists. Sextus's source for many of the ideas he outlines is Pyrrho (c. 360–c. 270 BCE) who, according to Diogenes Laertius, "held that there is nothing really existent, but custom and convention govern human action; for no single thing is in itself any more this than that." He showed his "indifference" to notions of truth, Diogenes writes, "by washing a porker." Another pig

---

[62] Julian Barnes, "Introduction" to Sextus Empiricus, in *Outlines of Scepticism*, ed. Julia Annas and Julian Barnes (Cambridge: Cambridge University Press, 2000), xii, xi. Boas mentions Sextus only once in *The Happy Beast*, and that in a footnote sourcing the story of Chrysippus's dog. Boas, *Happy Beast*, 7n15.

[63] Barnes, "Introduction" to Sextus Empiricus, xi.

[64] Montaigne, "Apology for Raymond Sebond," 477.

[65] Sextus Empiricus, *Outlines of Scepticism*, 3.

[66] Cicero, *The Academics of Cicero*, trans. James S. Reid (London: Macmillan, 1880), "Prior Academics," 60.

plays an important role in Pyrrho's thought, a pig referred to by Montaigne. Diogenes Laertius writes,

> When his fellow-passengers on board a ship were all unnerved by a storm, [Pyrrho] kept calm and confident, pointing to a little pig in the ship that went on eating, and telling them that such was the unperturbed state in which the wise man should keep himself.

Such, however, was Pyrrho's belief that nothing should be left to the "arbitrament of the senses," his friends had to follow him around to keep him "out of harm's way." They were obviously successful: he lived to be ninety.[67]

The pigs that figure in Pyrrho's life provide, in summary, illustrations of an important difference between academic and Pyrrhonic skepticism: in the work of an academic skeptic like Cicero, animals hardly feature; while in both Pyrrho's life, and in the fullest outline of his philosophy—Sextus's *Outlines*—animals are central. What is important is that Sextus constantly takes animals as evidence of the boundary of human understanding; he turns to what Jacques Derrida has termed the "abyssal limit of the human" in order to understand that human;[68] and by doing this, Sextus is doing something very different from Plutarch, something that Boas and his followers do not fully take into account in their discussion of the Theriophile tradition. In that tradition, as it is presented by Boas, animals are illustrations—Peter Harrison terms them "moral exemplars"[69]—that will lead humans to better lives. In skeptical thought they are no such thing.

The reason for the centrality of animals within Sextus's *Outlines* is because Pyrrhonic skepticism is a search for what can be truly known. In dogmatic philosophy—for example, the philosophy of Aristotle—one of the key things that is asserted to be known is the human, and it is from this human that other knowledge is built. In many ways some of the debates covered in the earlier chapters of this book reveal the break down in that dogmatic certainty and the continued (and often anxious) search for a foundational truth that is vanishing from view. What is central to Sextus's skepticism, and what has a bearing on the issues at stake in this book, is the question of the "standard by which"; that is, the basis of agreement by which proof itself can be judged. Who defines reason? and who then attributes it to one species—humanity—and not to any other? To ask other questions—how do humans know that animals do not experience prophetic dreams? What is the stan-

---

[67] Diogenes Laertius, *Lives of Eminent Philosophers*, trans. R. D. Hicks (London: William Heinemann, 1925), Volume II, Book IX, chapter II, "Pyrrho," pp. 475, 479, 481, 475.

[68] Jacques Derrida, "The Animal That Therefore I Am (More to Follow)," trans. David Wills, *Critical Inquiry* 28 (2002): 381.

[69] Harrison, "Virtues of Animals," 471.

dard that is used to judge the actions of the fox and to thus attribute them not to reason but to instinct? Or, to return to a certainty already discussed here: How does a human know whether a dog knows that it is a dog?

Charron had alerted his readers to the possibility of human partiality in the comparison of humans and animals, and this is a suggestion very much at one with Sextus's outline of skepticism. What is also important in Sextus's assessment is the possibility that it is human limitation, not human capacity, that creates the category "unreasonable." Sextus writes, "even if we do not understand the sounds of the so-called irrational animals, it is nevertheless not unlikely that they do converse and we do not understand them."[70] And Montaigne says, "It is no great wonder that we do not understand [animals]; neither do we understand the Basques and the Troglodytes."[71] It is not animals' failure to express themselves but our failure to comprehend them that is significant. Whereas, in the discourse of reason, language allowed humans to evidence their reasonable natures, and—by logic—animals' lack of language merely reveals their lack of reason, here it is that claim for animals' lack of language that is revealed to be unreasonable.

But, it would be too simple to assert that Sextus and skeptics following him argued for the rationality of animals. Finding no foundation from which to judge, apart from the limited and flawed human position, Sextus retires to what he terms "suspension of judgment," which "gets its name from the fact that the intellect is suspended so as neither to posit nor to reject anything because of the equipollence of the matters being investigated." This suspension of judgment is not the beginning of further enquiry, however; it is the end of enquiry and the beginning of "tranquility." And as Sextus states: "the causal principle of scepticism we say is the hope of becoming tranquil."[72] Wonder—enquiry, a turbulence of the mind—ceases, and peace reigns. This is the true aim of skepticism. In this way Pyrrhonic skepticism stands at odds with the concept of wonder that came from Aristotle by way of Albert the Great. Albert proposed that "wonder is the movement of the man who does not know on his way to finding out, to get to the bottom of that at which he wonders and to determine its cause. . . . Such is the origin of philosophy."[73] Skeptical wonder is an acknowledgement of irresolution, or of the impossibility of knowing, and as such is the beginning of calmness. What begins philosophy for one thinker signals its end for another.

Sextus's use of animals as the limit case of human intellect is a trope that

[70] Sextus Empiricus, *Outlines of Scepticism*, 22.

[71] Montaigne, "Apology for Raymond Sebond," 402.

[72] Sextus Empiricus, *Outlines of Scepticism*, 73–74, 75–76, 49, and 5.

[73] Albertus Magnus, *Commentary on the Metaphysics of Aristotle*, cited in *Marvelous Possessions: The Wonder of the New World*, by Stephen Greenblatt (Oxford: Clarendon Press, 1991), 81.

is, as we have seen, repeated by Montaigne in the "Apology for Raymond Sebond" and by Charron in *Of Wisdome*. But it is not only in France that Sextus's influence can be felt. A clear adherence to skepticism can be found in the English text *The Sceptick* (c. 1590), which was attributed to Sir Walter Raleigh when it was first published in 1651. Here the focus is almost entirely on animals, and the questions asked are wholly conventional. As William M. Hamlin has noted, this English text "loosely translates" parts of the *Outlines*.[74] In this short text (Hamlin's modern edition takes up only ten pages) the author makes the rationale for the focus on animals clear: the "first reason" for skepticism, he argues, "ariseth from the consideration of the great difference amongst living creatures, both in the matter and manner of their generations, and the several constitutions of their bodies." The acknowledgment of natural differences of "temperament and quality" causes, so the author affirms, "a great diversity in their fantasy and conceit." But the skeptical assertion of difference does not allow for an assertion of human superiority; far from it. The author of *The Sceptick* turns to the organs of perception and notes that if the organs of perception are different in animals and in humans (as, he asserts, they are), then it is only to be expected that what is apprehended by those organs also differs: "it is very probable that fishes, men, lions, and dogs, whose eyes so much differ, do not conceive the self same object after the same manner, but diversely according to the diversity of the eye, which offereth it unto the fantasy." The same, the author argues, holds true for all the senses, and leads to a series of telling questions: "why should I presume to prefer my conceit and imagination in affirming that a thing is thus and thus in its own nature, because it seemeth to me to be so, before the conceit of other living creatures, who may as well think it to be otherwise in its own nature, because it appeareth otherwise to them than it doth to me?"; "why then should I condemn [animals'] conceit and fantasy concerning any thing more than they may mine?"[75] The author offers no clear answers to these questions; he is a skeptic, and as Jonathan Barnes has noted, "The Sceptical investigator neither asserts nor denies, neither believes nor disbelieves."[76] The questions are asked in order to mark out the limits of human reason; to indicate the ways in which what is claimed for human understanding goes beyond what is provable and moves into the realms of arrogance, self-deceit, and, we might add, anthropocentrism. What is significant is that it is through animals that such a limit is marked, because

---

[74] William M. Hamlin, "A Lost Translation Found? An Edition of *The Sceptick* (c.1590) Based on Extant Manuscripts [with text]," *English Literary Renaissance* 12, no. 2 (2001): 34. I use Hamlin's edition throughout.

[75] *The Sceptick*, 42, 43.

[76] Jonathan Barnes, "The Beliefs of a Pyrrhonist," *Proceedings of the Cambridge Philological Society* 208, n.s., 28 (1982): 1.

it is in animals that human partiality is most explicitly illustrated. Animals mark the audacity and the limit of the human.

The point that follows the second question asked by the author of *The Sceptick*—"why then should I condemn [animals'] conceit and fantasy concerning any thing more than they may mine?"—is as close as he is able to get to an answer: "They may be in the truth and I in error, as well as I in the truth and they err." It is a hypothetical situation to which no conclusion is offered, and the lack of a conclusion leads to an undermining of human status and a repetition of Sextus:

> If my conceit must be believed before theirs, great reason that it be proved to be truer than theirs. And this proof must be either by demonstration or without it. Without it none will believe, certainly. If by demonstration, then this demonstration must seem to be true, or not seem to be true. If it seem not to be true, it is easily rejected. If it seem to be true, then will it be a question whether it be so indeed as it seemeth to be. And to allege that for a certain proof which is uncertain and questionable seemeth absurd.[77]

How can we *prove* human preeminence? In a sense, this question had not been asked within the discourse of reason. There the assertion of the a priori existence of the rational soul was taken as proof enough. In the millennia following Aristotle it was an orthodoxy of unthought proportions that man *was* preeminent; that was the starting point from which all thinking emerged. Skeptical thinkers do not accept the unthought assertion of the truth of human judgment as the "standard by which" judgment itself is made. The truth of human judgment is a presupposition that needs to be inquired into, wondered at. And because inevitably it cannot be answered without recourse to human judgment, it is impossible to uphold. The final words of *The Sceptick* represent a finality that is forever deferred: "I may then report how these things appear, but whether they are so indeed, I know not."[78]

Thus within skepticism humans do not have to establish that animals are superior to them, as Boas's and others' understanding of the Theriophile tradition holds. Rather, humans have to think about the ways in which their epistemological constructions work, about the ways in which they, humans, structure the world around them. This epistemological inquiry does not, though, remove the work of skeptical thinkers from real animals (although in skeptical terms, of course, the concept of the "real" is clearly problematic). As Katherine Eisaman Maus has noted, their "perspectivism seems to

---

[77] *The Sceptick*, 45.
[78] Ibid., 51.

strengthen, not weaken, the impulse to investigate those [animal] minds."[79] To posit that a dog might also structure the world around it is to take note of empirical as well as epistemological data. And such a proposal also offers an opportunity to take the debates about the existence in humans of the rational soul and see them for what they are: debates constructed by humans for the benefit of humans. As Charron noted, this does seem rather "partial."

Montaigne's response, echoed by some English thinkers, is to prioritize sentience, something that both animals and humans possess, and to acknowledge that human superiority may be a myth.

> Let [man] help me to understand, by the force of his reason, on what foundations he has built these great advantages that he thinks he has over other creatures. . . . And this privilege that he attributes to himself of being the only one in this great edifice who has the capacity to recognize its beauty and its parts, the only one who can give thanks for it to the architect and keep an account of the receipts and expenses of the world: who has sealed him this privilege? Let him show us his letters patent for this great and splendid charge.[80]

The assertion made by Joseph Hall, quoted earlier, that only humans can truly "view" and think about the world has here been undone. Such human capacity is revealed not as truth but as opinion; not as reflective of a natural hierarchy but as constitutive of that hierarchy. To ask as Montaigne does for the "letters patent" of human superiority—to attempt to find the origin of the invention of human "privilege"—is to undercut the naturalness of human dominion, to view reason not as the true marker of humanity but as a creation of humans for humans.

In the representation of animals in early modern thought, then, what can be traced is a parallel to some of the debates about the nature of the human. Animals are potentially reasoning—such can be witnessed by observing their actions and by the application of a different understanding of the seat of reason from that offered by the discourse of reason. But animals are also unreasonable, are objects for use, and are given meaning through preexisting narratives. As well as this, and more troubling than this, is the skeptical assertion that the categories used to establish the nature of humans and animals are themselves worth questioning: that the notion "reason," for example, might not reflect a given issue, but might be used to constitute an ideal from which difference is constructed.

---

[79] Katherine Eisaman Maus, *Inwardness and Theater in the English Renaissance* (Chicago: University of Chicago Press, 1995), 7.

[80] Montaigne, "Apology for Raymond Sebond," 399.

The fact that Sextus Empiricus's writings are first printed in the late sixteenth century is, perhaps, both luck—the discovery of various manuscripts earlier in the sixteenth century obviously led the way—and shows a sense of their growing relevance to contemporary readers. Richard H. Popkin has argued that skepticism becomes visible at the beginning of the Reformation: "The problem," he writes, "of justifying a standard of true knowledge does not arise as long as there is an unchallenged criterion. . . . The Pandora's box that Luther opened at Leipzig was to have the most far-reaching consequences, not just in theology but throughout man's entire intellectual realm."[81] Among the consequences, I want to argue, was a revised assessment of the human perceptions of animals.

It is not going too far, perhaps, to argue that the reemergence of skepticism offers writers a way of thinking through the various logical breakdowns in the discourse of reason. No longer easy to sideline, the collapse of certainty—of human superiority and animal inferiority, of the opposition of reason and unreason—became a focus of anxiety and, for a few, celebration. Animals emerge, to put it simply, as many things to many people. They are objects of human use and subjects of human discussion; they are beyond the human and are the limitation of the human; they are ideal and they are real. In early modern England there were traditions to support all of these different perspectives.

But even in the face of such challenges the myth of human superiority was, of course, a powerful one, and it is possible to see the desire for the maintenance of this myth at work in writings surrounding what was, perhaps, the most famous "reasonable" animal in early modern England: Morocco the Intelligent Horse. However, as well as tracing orthodox arguments about human superiority in discussions of this equine wonder, we may also observe the appearance of some of the ideas about animals that arise from within the skeptical tradition. It may seem to be a leap to move from a discussion of the difference between Plutarch and Sextus Empiricus, between Academic and Pyrrhonic skepticism, to debates about a performing horse, but actually, as the plethora of writings about Morocco from early modern England reveals, this horse was a figure not merely of entertainment and comedy but of philosophy as well. To think about Morocco is to think, in fact, about thinking.

[81] Richard H. Popkin, *The History of Scepticism from Erasmus to Descartes* (New York: Harper Torchbook, 1968), 4.

CHAPTER 5

A Reasonable Animal?

In 1609 Thomas Morton, the Bishop of Chester, recalled a story he had been told about Morocco the Intelligent Horse. The owner of the horse, Bankes, related

> his own experience in *France* among the Capuchins, by whom he was brought into suspition of Magicke, because of the strange feats which his horse *Morocco* plaied . . . at *Orleance:* where he, to redeeme his credit, promised to manifest to the world that his horse was nothing lesse then [i.e. not] a Diuell. To this end he commanded his horse to seeke out one in the preasse of the people, who had a crucifixe on his hat; which done, he bad him kneele downe vnto it; & not this onely, but also to rise vp againe, and to kisse it. And now (Gentlemen quoth he) I thinke my horse hath acquitted both me, and himselfe; and so his Aduersaries rested satisfied: conceauing (as it might seeme) that the Diuell had no power to come neare the Crosse.[1]

Morton tells this tale not in admiration of the capacity of the beast but in order to mock the Capuchins. He is attacking the idea of the efficacy of the cross—the belief in its power to move—and the horse is used to do this. The question that Morton raises is: Which is more ignorant, to imagine that there is such power in a mere object, or to believe that an animal can be inspired by the Holy Spirit? It is difficult, the story implies, to work out which is worse.

This tale about Morocco is in many ways typical of the early modern pe-

---

[1] [Thomas Morton], *A Direct Answer Unto the Scandalous Exceptions which Theophilus Higgons hath lately objected against D. Morton* (London, 1609), 11.

riod. The animal is used to discuss issues in the human domain, and because of this the animal as animal vanishes. But to merely end the story here, to only note the way in which Morocco was used to speak about other issues, would be a failure to recognize that Morocco the Intelligent Horse was of interest in himself and to notice that references to him appear in numerous texts in the late sixteenth and early seventeenth centuries: in religious works and social commentaries; in plays, poetry, pamphlets, almanacs; in margins. The sheer volume of references would seem to imply that Morocco was infamous in England, and his infamy is tied up with debates that are central to this book.

A Horse's Life

Reports of Morocco and his master Bankes date from 1590, when Sir John Davies noted that even by then "Bankes his horse [was] better knowne" than he.[2] The horse's performances contained a number of particular elements. Morocco's ability to pick out individual members of the audience is mentioned in a number of texts: in *Haue with you to Saffron-walden* Thomas Nashe wrote that "Bankes his horse knowes a Spaniard from an Englishman." Almost twenty years later Richard Brathwaite noted his ability "To know an honest woman from a whoore."[3] Morocco also had other talents: in 1598 Thomas Bastard wrote "*Bankes* hath an horse of wondrous qualitie, / For he can fight, and pisse, and daunce, and lie. / And finde your purse, and tell what coyne ye haue."[4] George Peele even writes of himself that "hee had borrowed a Lute to passe away the melancholy afernoone, of which he could play as well as *Bankes* his horse."[5] The ridiculousness of the image—which of course works to reveal Peele's musical abilities to be severely limited— also reveals the popular perception of Morocco as a performer. But, while he might not be able to play the lute, Morocco could, according to the only known image of him, count the spots on dice (figure 5.1).

This image comes from a satirical pamphlet of 1595, and according to its authors, the pseudonymous John Dando and Harry Runt, Morocco could even discourse with his owner about the hypocrisy of sadlers, whores, landlords, and gallants. The pamphlet is apparently a report of an overheard

[2] John Davies, "Ad Musam 48," in *Epigrammes* (Middleburgh, ?1590), sig. D3ᵛ.

[3] Thomas Nashe, *Haue with you to Saffron-walden* (1596), cited in J. O. Halliwell-Phillips, "The Dancing Horse," in *Memoranda on Love's Labour's Lost, King John, Othello, and on Romeo and Juliet* (London: James Evan Adlard, 1879), 52; Richard Brathwaite, *A Strappado for the Divell* (London: Richard Redmer, 1615), 159. Because of the apparent length of his career, it is likely that more than one horse was trained to perform as Morocco.

[4] T. B. [Thomas Bastard], "Epigr. 17. Of Bankes horse," in *Chrestoleros* (London: I. B., 1598), 62.

[5] George Peele, *Merrie Conceited Iests* (London: F. Faulkner, 1627), 4.

Figure 5.1. Morocco the Intelligent Horse, from John Dando and Harry Runt, *Maroccus Extaticus. Or, Bankes Bay Horse in a Trance* (London: Cuthbert Burby, 1595). By permission of the British Library, classmark C40c29.

conversation between Morocco and Bankes; the authors comment "for truly there was neuer horse in this world aunswered man with more reason, nor neuer man in this world reasond more sensibly with a horse than this man and this horse in this matter." Morocco himself comments, "this Latine I learned when I gambolde at Oxforde . . . *O tempora, O mores, O Poetarum flores.*"[6] The line comes from Cicero, and Morocco's assertion that his atten-

[6] John Dando and Harry Runt, "To the Reader," in *Maroccus Extaticus. Or, Bankes Bay Horse in a Trance* (London: Cuthbert Burby, 1595), n.p. and sig. B3ᵛ.

dance at the great seat of learning made him literate (and thus human) is echoed in a couplet quoted by Owen Feltham: "A thing borne blinde, a child, and foolish too, / Shall be made a man, if it to *Oxford* goe."[7] It therefore seems hardly surprising to find, in 1618, Sir John Harington interpreting a comparison made between his own dog and Morocco in favorable terms.

> The Dogge is grac't, compared with great Bankes,
> Both beasts right famous, for their pretty prankes,
> Although in this, I grant, the dogge was worse,
> He onely fed my pleasure, not my purse.[8]

Morocco's capacity to earn his owner a good living is what places him above Harington's dog.

But it was not only London residents who were able to see Morocco's performances. A diary entry from September 1591 records a provincial performance: "This yeare and against the assise tyme on Master Banckes, a Staffordshire gentile, brought into this towne of Salop a white horsse whiche wolld doe woonderfull and strange thinges."[9] But Morocco was not confined to England; Jean de Montylard's 1602 French translation of Apuleius's *The Golden Ass* contains a marginal note which records the appearance in Paris of "Le cheval . . . Moraco," and in 1609 Thomas Morton, as noted above, reports the horse's appearances in Orleans and Frankfurt.[10]

De Montylard is not alone in linking Morocco with classical precedent when he mentions the horse in his translation of *The Golden Ass;* Richard Sorabji labels Apuleius, like Plutarch, a "Middle Platonist,"[11] and in 1615 Henry Peacham's epigram, "*Asinas ex Asino*," presents "*Grillus*. . . Dreaming of late hee was transformd an Asse" and attempting "eke like *Bankes his Horse* to daunce."[12] The transformed man of classical myth becomes, it would seem, the intelligent horse of contemporary Europe.

Horse and man are linked to another figure from classical history in 1610, when Ben Jonson records the tragedy of "Old Banks the juggler, our

---

[7] Owen Feltham, *Resolves or, Excogitations. A Second Centurie* (London: Henry Seile, 1628), 161.

[8] John Harington, "Against Momus, in praise of his dogge Bungey," in *The Most Elegant and Witty Epigrams of Sir John Harington, Knight* (London: Iohn Budge, 1618), sig. I[v].

[9] MS cited in William Shakespeare, *Love's Labour's Lost* (New Variorum Edition), ed. Horace Howard Furness, 4th impression (Philadelphia: J. B. Lippincott, 1904), 45 ad 1.2.50.

[10] Jean de Montylard, cited in Halliwell-Phillips, *Memoranda on Love's Labour's Lost*, 31; [Morton], *Direct Answer*, 11.

[11] Sorabji, *Animal Minds*, 220, 233.

[12] H. P. [Henry Peacham], "*Asinas ex Asino*," in *The Mastive, or Young-Whelpe of the Olde-Dogge* (London: Tho. Creede, 1615), sig. G3[v].

Pythagoras, / Grave tutor to the learned horse. Both which, / Being, beyond sea, burned for one witch."[13] Jonson's reference to Pythagoras refers, perhaps, to that philosopher's belief in the transmigration of souls. In Arthur Golding's translation of Ovid this belief is outlined:

> This same spright [soul]
> Dooth fleete, and fisking heere and there dooth swiftly take his flyght
> From one place too another place, and entreth euery wyght,
> Remouing out of man too beast, and out of beast too man.
> But yit it neuer perrisheth nor neuer perrish can.[14]

In his image of Bankes as "our Pythagoras," then, Jonson may be comically representing Morocco as evidence of the human soul having entered the animal body (although, as quoted below, Sir Walter Raleigh offers a different explanation of the link to Pythagoras). The dangerous possibility of the transmigration of souls—for what would happen to human superiority in such a situation?—is one that Golding himself defends against in the "Epistle to Lord Robert, Earl of Leicester" that precedes his translation of the *Metamorphoses:*

> But as for that opinion which Pythagoras there brings
> Of soules remouing out of beasts too men, and out of men
> Too birdes and beasts both wyld and tame, both too and fro agen:
> It is not too be vnderstand of that same soule whereby
> Wee are endewd with reason and discretion from on hie:
> But of that soule or lyfe the which brute beasts as well as wee
> Enioy.[15]

However, Jonson's poem records the tragic news that both man and horse died together "beyond sea," and further information about this is given in the proto-novel *Don Zara Del Fogo: A Mock-Romance* (1656). A marginal note, associated with the description of Zara's horse as being "like one of *Banks* breed," reads:

> Meaning Banks his Beast, if it be lawful to call him a beast, whose perfections were so incomparably rare, that he was worthily termed the four-legged won-

[13] Ben Jonson, "On the Famous Voyage," in *Ben Jonson: The Complete Poems,* ed. George Parfitt (London: Penguin, 1988), 91.
[14] Ovid, *The XV Bookes of P. Ouidius Naso, entytuled Metamorphosis,* trans. Arthur Golding (London: Willyam Seres, 1567), Book XV, p. 382.
[15] Arthur Golding, "The Epistle to Lord Robert, Earl of Leicester," in *Metamorphosis,* sig. aii^r.

der of the world, for dancing, (some say) singing, and discerning Maids from Maulkins, finally having of a long time proved himself the ornament of the Brittish Clime, travailing to Rome with his Master, they were both burned by the commandment of the Pope.[16]

This story of the demise of the horse and his master is not necessarily true—references after 1610 to one Bankes, a vintner in Cheapside, recorded by Halliwell-Philips, might refer to the owner of Morocco,[17] and numerous post-1610 texts continue to invoke Morocco, implying his continuing popularity. But the important fact when reading Jonson's reference is that the story of human's and horse's death was told, and that it was told to an audience who, we must assume, were expected, if not to believe it, then at least to recognize it as possible; that their burning together as one witch was a potential response to Morocco's performances. The reason why such a fate was considered possible is contained in one of the early modern interpretations of the intelligent horse, and in this explanation, and the resistance to it, the problems which Morocco's feats raise start to become clear: How can one maintain human superiority—maintain, in fact, the discourse of reason—in the face of an intelligent horse?

Unnatural Acts

The problem is a simple one: if an animal is incapable of abstract reasoning, as so many writers from classical times onwards argue, then an animal cannot display the abilities of counting, discerning, judging; the two are incompatible. One solution, therefore, was to regard the horse as unnatural. But the possibility that Morocco, in displaying these abilities, was more than a horse—that he was not, in fact, natural—rebounded inevitably on the activities of his owner. The Salop resident who recorded Morocco's performance in 1591 wrote that "many people judgid that it were impossible to be don except he had a famyliar or don by the arte of magicke."[18] Bankes, it was proposed, was a magician; and Morocco, not natural but supernatural.

Offering a different interpretation of the link between Pythagoras and Bankes, and emphasizing Iamblichus's testimony that Pythagoras was able

---

[16] *Don Zara Del Fogo: A Mock-Romance* (London: Tho. Vere, 1656), 114.

[17] Halliwell-Phillips records a "Bill of Fare, sent to Bankes the Vintner in Cheapeside, in May, 1637," in which the human–animal boundary that is problematized in Morocco's act is parodied: the menu includes "Foure paire of elephants' pettitoes . . . a rhinoceros boyled in alligant; sixe tame lyons in greene sawce. . . ." Cited in Halliwell-Phillips, *Memoranda on Love's Labour's Lost,* 52.

[18] MS cited in Shakespeare, *Love's Labour's Lost,* ed. Furness, 45.

to speak to animals,[19] Sir Walter Raleigh wrote of the power "which they call charming of Beasts and Birds, of which *Pythagorus* was accused," and argued that "certainly if *Banks* had liued in elder times, hee would haue shamed all the Inchanters of the World: for whosoeuer was most famous among them, could neuer master or instruct any Beasts as he did his Horse." Raleigh calls this ability to train animals deceit or "pettie Witchery."[20] This explanation does solve the problem of the intelligent horse in one way—it means that a natural horse can never be intelligent, thus maintaining a clear boundary between humans and animals—but it also opens up another problem.

For a human to transform an animal by magic into an intelligent beast— to turn, in Morocco's case, one of the most useful creatures to man into a threat to human status—is not to extend human power over the natural world. Rather, it is the opposite; it is to undo the dominion that God had given to Adam. If Morocco can see through the surface—can tell a virgin from a whore—he is replicating Adam's naming of the beasts (Genesis 1:19). As Thomas Adams wrote: "Hee at first saw all their insides."[21] This is exactly what Morocco seems to be capable of.

But the theological implications of the magical explanation are broader than this. In *The discoverie of witchcraft* (1584) Reginald Scot wrote of the blasphemy of "attributing vnto others, that power which onelie apperteineth to God, who onelie is the Creator of all things."[22] From this perspective, any act that is performed has to come through the will of God, and to claim that humans or the devil could direct events is to overstep the bounds of righteousness. Scot writes, "he that attributeth to a witch, such diuine power, as dulie and onelie apperteineth vnto GOD (which all witchmongers doo) is in hart a blasphemer, an idolater, and full of grosse impietie, although he neither go nor send to hir for assistance."[23] If we take up this analysis of sorcery here, it has to be acknowledged that attributing magical powers to Bankes is blasphemous; it is claiming for the man a power that is only God's. By offering a supernatural explanation for what apparently exceeds nature, writers are, then, creating a new problem for themselves: the human— Bankes—becomes more powerful than God. This would overturn the placement of humans *between* angels and animals that is proposed in the discourse of reason. But as well as this theological worry, there also lurks a possibility

[19] Iamblichus, *Iamblichus' Life of Pythagoras, or Pythagoric Life*, trans. Thomas Taylor (1818; repr., London: John M. Watkins, 1926), 29–30.

[20] Sir Walter Raleigh, *The History of the World. In Five Bookes* (London: Walter Burre, 1614), 178.

[21] Thomas Adams, *Meditations Vpon Some Part of the Creed*, in *The Works of Tho: Adams* (London: Tho. Harber, 1630), 1132.

[22] Reginald Scot, *The discoverie of witchcraft, Wherein the lewde dealing of witches and witchmongers is notablie detected* (London: W. Brome, 1584), sig. Aiij[r].

[23] Scot, *The discoverie*, 12.

that could endanger all human–animal relations. If an animal (even by magic) can gain access to that which is human—namely, reason—then surely the boundary that separates man from beast has already collapsed. This is why Jonson writes that Bankes and Morocco die as "one witch": the distinction of owner and owned, master and servant, human and animal, has gone.

Later in *The discoverie*, Reginald Scot emphasizes this collapsing of the boundaries in a very different way. Scot relates a story about an ass that comes originally from two of the sources of the belief in witches that he is attacking in his work: Jakob Sprenger and Heinrich Kramer's *Malleus maleficarum* (1486) and Jean Bodin's *De la démonomanie des sorciers* (1580). The story Scot retells "happened in the citie of *Salamin,* in the kingdome of *Cyprus*" and is strangely similar in some ways to Thomas Morton's later story of Morocco in Orleans. In Scot's tale an English sailor has shore leave and goes "to a womans house, a little waie out of the citie, and not farre from the sea side, to see whether she had anie egs to sell." The woman is, of course, a witch, who decides "to destroie him"; she gives him some eggs and mysteriously tells him "to returne to hir, if his ship were gone when he came."

> The yoong fellowe returned towards his ship: but before he went aboord, hee would needs eate an eg or twaine to satisfie his hunger, and within short space he became dumb and out of his wits (as he afterwards said.) When he would haue entred into the ship, the mariners beat him backe with a cudgell, saieng; What a murren lacks the asse? Whither the diuell will this asse? The asse or yoong man (I cannot tell by which name I should terme him) being many times repelled, and vnderstanding their words that called him asse, considering that he could speake neuer a word, and yet could vnderstand euerie bodie; he thought that he was bewitched by the woman, at whose house he was.

The man has been transformed into an ass, and he returns to the woman who gave him the eggs "in whose seruice hee remained by the space of three yeares, dooing nothing with his hands all that while, but carried such burthens as she laied on his backe." Then, one Sunday morning, the ass found himself alone before the church. He

> heard a little saccaring bell ring to the eleuation of a morrowe masse, and not daring to go into the church, least he should haue beene beaten and driuen out with cudgels, in great deuotion he fell downe in the churchyard, vpon the knees of his hinder legs, and did lift his forefeet ouer his head, as the preest doth hold the sacrament at the eleuation. Which prodigious sight when certeine merchants of *Genua* espied, and with woonder beheld; anon commeth the witch with a cudgell in hir hand, beating foorth the asse. And

bicause (as it hath beene said) such kinds of witchcrafts are verie vsuall in those parts; the merchants aforesaid made such meanes, as both the asse and the witch were attached by the iudge. And she being examined and set vpon the racke, confessed the whole matter, and promised, that if she might haue libertie to go home, she would restore him to his old shape: and being dismissed, she did accordinglie. So as notwithstanding they apprehended hir againe, and burned hir: and the yoong man returned into his countrie with a ioifull and merrie hart.[24]

In this story, as in the Capuchins' interpretation of Morocco, it is the worship of religion that reveals the truth. But unlike Morton's tale, where, for the Capuchins, the naturalness of Morocco is proved by his reverence of the cross, the tale Scot tells of the ass apparently proves the opposite: faith, the story says, is confined to humanity, and an animal that worships God is clearly not natural. What is clear is that in this tale the human remains constant, even in the body of the ass.

Scot's telling of the story of the transformed human is a version of a longer tale told by Apuleius in *The Golden Asse,* in which the author records his time in the shape of an ass, following his taking of a magical potion. An English translation of this Latin text was published in 1566, going through five further editions before 1639.[25] Like the English edition of *The Golden Asse,* the story Scot tells makes no direct reference to Morocco, but in two later versions of a very similar tale of magical transformation, the intelligent horse is invoked, thus reinforcing the sense that Morocco and magical transformation were linked in the minds of some early modern writers. In his *Gynaikeion* (1624) Thomas Heywood relates the following tale in a chapter devoted to "Witches that haue eyther changed their owne *shapes, or transformed others.*"

A Minstrell or Pyper trauelling [near Rome], tasted of this cheese and was presently changed into an Asse, who notwithstanding hee had lost his shape, still retained his naturall reason, and (as one Bankes here about this citie taught his Horse to show trickes, by which he got much money) so this Asse being capable of what was taught him, and vnderstanding what he was bid to doe, showed a thousand seuerall pleasures (almost impossible to be apprehended by any vnreasonable creature) to all such as came to see him and payde for the sight.[26]

---

[24] Scot, *The discouerie,* 94–95 and 95–96.

[25] The translation, by William Adlington, was titled *The xi Bookes of Golden Asse, conteininge the Metamorphosie of Lucius Apuleius* (London: Henry Wykes, 1566).

[26] Thomas Heywood, *Gynaikeion: or, Nine Bookes of Various History Concerninge Women; Inscribed by ŷ name of ŷ Nine Muses* (London: Adam Islip, 1624), 411.

Heywood then goes on to tell the same story of the English sailor as Scot does. Nine years later William Prynne, the great antitheatricalist, related the story of "*a certaine Stage-player whot got his living by acting.*" This player, through the sorcery of the two witches he lives with, is "*metamorphosed into the shape of an Asse.*" Like Heywood before him, Prynne has a contemporary image to emphasize the nature of the transformation: "*being thus transformed, he became so tractable that* (like another *Bankes* his dancing Horse, or the *dancing Horses of the Sybarites and Cardians) he would readily turne and move which way soever these Witches commanded him.*"[27] For Prynne, the horse has the same status as the actor; he is incapable (paradoxically) of acting, but can only suffer, can only ventriloquize. Prynne concludes:

> If this bee but an *Ovids Metamorphosis,* or an *Apuleius his Golden Asse;* we may laugh at the conceit, and so passe it by: but if it bee a truth, as the Historian confidently affirmes it, wee may deeme it a just judgement of God upon this Actor, who for his acting of other mens parts in jest, was thus enforced to play the Asses part in earnest.

Heywood and Prynne, for very different reasons, both turn to stories of human transformation, and both recognize a contemporary parallel that might be of help in their storytelling. Morocco is a horse and he is more than a horse. There are fictional forebears—Apuleius and Ovid—but there may also be factual—historical—ones.

However, in neither Heywood's nor Prynne's invocations of Morocco is the horse the focus, and like these two later writers Reginald Scot also tells his story of the transformed sailor not in order to address the issue of the status of beasts but to make a specific theological point. His refusal to believe the tale of the transformed sailor has a simple foundation: "I woonder," he writes, "at the miracle of transubstantiation." He wonders further "that the asse had no more wit than to kneele downe and hold vp his forefeete to a peece of starch or flowre, which neither would, nor could, nor did helpe him." It is, as he points out, the witch herself—not God—who undoes the spell. Attacking the support given to the fable of the ass by Jean Bodin and the authors of *Malleus maleficarum*, Scot asks "But where was the yoong mans owne shape all these three yeares, wherein he was made an asse? It is certeine and a generall rule, that two substantiall formes cannot be in one subiect *Simul & semel,* both at once." His conclusion is simple: "the whole storie is an impious fable."[28] Just as bread is bread and the body of Christ is the body of Christ, a human, he implies, is a human, and an animal an animal. There

---

[27] William Prynne, *Histrio-Mastix* (London: Michael Sparke, 1633), 553.
[28] Scot, *The discoverie*, 96.

is no point at which both can exist together. In another tale Scot relates of a performing animal—one who fakes death, shows emotions, and responds to language—he is equally clear: "*Bodin* saith, that this was a man in the likenesse of an asse: but I maie rather thinke that he is an asse in the likenesse of a man."[29] Human reason, of course, can be lost—witness Bodin's foolish belief in transformation— but an animal can never have reason.

In this way, supernatural explanations for Morocco's tricks, while they offer fictional precedent for the horse, do not work to reestablish human superiority because they open up as many problems as they appear to shut down. They not only imply that the intelligent animal is evidence of human blasphemy (only God has the power to transform) but also offer up the frightening possibility that the boundary between human and animal is not fixed. Another explanation for the intelligent horse must be sought, then, and an alternative account is to be found in nature, not in supernature. But the natural explanation, like so many others we have examined, contains the seeds of its own undoing. An important question is being asked—How limited are the capacities found in the organic soul of an animal?

Scholarship

The discoverie of witchcraft was written for a specific purpose: Scot states, "*My question is not (as manie fondlie suppose) whether there be witches or naie: but whether they can doo such miraculous works as are imputed vnto them.*" His intention was to "*See first whether the euidence be not friuolous, & whether the proofs brought against them be not incredible, consisting of ghesses, presumptions, & impossibilities contrarie to reason, scripture and nature.*"[30] A similar intention can be traced in the last chapter of a very different text, Gervase Markham's *Cavelarice, Or the English Horseman* (1607), which turns ultimately to the most obvious illustration of equine intelligence available at the time. It is when he looks at Morocco that Markham is able to truly explain how this horse's apparent intelligence works on the English stage. It is, Markham argues, "generall opinion . . . that it was not possible to bee done by a Horse, that which that Curtall did, but by the assistance of the Deuill." But just as Scot intends to undercut arguments about the power of witches, so Markham makes it his job to prove that Bankes is "exceeding honest," and "that ther was no one tricke which that Curtall did, which I will not almost make any Horse do in lesse then a months practise."[31] Whereas Scot relies on theology to make his

---

[29] Ibid., 253–54.

[30] Ibid., sigs. A8ᵛ and A6ʳ.

[31] Gervase Markham, *Cavelarice, Or the English Horseman* (London: Edward White, 1607), VIII, pp. 26–27.

point—to turn Bodin into an ass—Markham relies on whips and bridles to turn any horse into Morocco. We shift, in fact, from a supernatural to a natural explanation for Morocco's performance.

This is made clear in Markham's outline of how to train a horse to count. It is worth quoting at length:

> Now, if you will teach your Horse to reckon any number, by lifting vp and pawing with his feete, you shall first with your rodde, by rapping him vpon the shin, make him take his foote from the ground, and by adding to your rod one certaine word as *Vp:* or such like, now whē he will take vp his foote once, you shall cherrish him, & giue him Bread, and when hee sets it vppon the ground, the first time you shall euer say one, then giue him more bread, and after a little pause, labour him againe at euery motiō, giuing him a bit of bread til he be so perfit, that as you lift vp your rod, so he will lift vp his foot, and as you moue your rod downeward, so he will moue his foot to the ground, and you shall carefully obserue to make him in any wise to keep true time with your rod, and not to moue his foot when you leaue to moue your rodde, which correcting him when he offends, both with stroakes and hunger, he will soone be carefull to obserue, after you haue brought him to this perfectnesse, then you shall make him encrease his numbers at your pleasure. . . .[32]

Two things immediately come to the fore in Markham's training methods: the cruelty to the horse, which can only be perceived if one recognizes a clear organic similarity between human and animal (thereby recognizing that the horse can feel the pain); and, at the same time, the crucial difference between them, by which the right of violence is given to the trainer.

Markham's position on this is utterly conventional. In his *Fower chiefyst offices belongyng to Horsemanshippe* (1565), based on Frederico Grisone's 1550 Italian original, Thomas Blundeville provided an English-speaking readership, and future generations of authors of horse training manuals, with an orthodox outline of training and of the nature of the horse. The rest between training sessions, Blundeville argues, should not be too long in case it "doth cause [the horse] to forgette his lessons." And the training should happen "in one place" so that the horse "shall better remember his lessons: yea and also his corrections, whereby there shall be bredde in him as it were an habit of well doyng." But such recognition of the limited (organic) capacity of the horse to remember (the event must be recently impressed on his memory to still be there; he must be in the same place for all lessons) is not always enough, and Blundeville—taking his lead from Grisone—suggests some astonishing means of bringing the horse into line:

[32] Ibid., VIII, pp. 33–34.

Let a footeman stande behinde you with a shrewed cat teyed at the one ende
of a long pole with her belly vpward, so as she may haue her mouth and
clawes at libertye. And when your horse doth stay or go backward, let him
thrust the Catte betwixt his thighes, somtime by the rompe, and often times
by the stones.

He follows this by suggesting similar torments using a hedgehog and a
"whelpe" (puppy). But such torments as these, he adds, are "not to be commonlye vsed but onelye in time of neede, and that with grat discretion."[33]
Blundeville's horse-training manual, however unreasonable it might appear
to us, clearly operates within the discourse of reason. Even as it shows how
to transform a horse, it shows that horse possessing only organic capacities:
it can remember only in a limited manner and, crucially, can be physically
tormented (that is, can feel pain) and from that experience of pain learn
obedience.

All this is followed by Markham in *Cavelarice*. Blundeville, for example,
recommends the use of the "sound of the voyce" in training, with tone being the crucial signifier of meaning to the horse:

if you would correct him for anye shrewde toye or obstinacye, you must al to
rate him with a terrible voyce, saying to him. Ah traitor, Ah villain, tournc
here, stop there, and suche like. But yf you woulde helpe hym at anye time,
then you must vse a more mylde and chearfull voyce, as when you run hym,
to say hey, hey, or now now.[34]

Markham likewise urges his readers to speak to their horse during training,
and in doing so reveals his indebtedness to the earlier writer:

If your Horse out of ignorance bee about to doe contrary to your will, then
to vse this word, *Be wise*, at which if he do not stay and take better deliberation, but wilfully pursue his error, then to correct him and vse this word *Vil-
layne* or *Traitor*, or such like, so you vse but one word; and when he doth as
you woulde haue him, to cherrish him, and vse this word *So boy*, in a short
space you shal bringe him to that knowledge that he wil wholy be directed
by those words and your commaundement.[35]

This advocating of a conversation with the horse does not imply the horse
possesses the capacity to understand language. Albert the Great, for exam-

---

[33] Thomas Blundeville, *The Fower chiefyst offices belongyng to Horsemanshippe* (London: William Serres, 1565), "Arte of Rydinge," fols. 17ᵛ–18ʳ, 19ᵛ–20ʳ, and 63ᵛ.

[34] Ibid., fols. 5ʳ and 8ʳ⁻ᵛ.

[35] Markham, *Cavelarice*, VIII, pp. 29–30.

ple, argued that animals lack language. Following classical convention, he distinguished between *sonus, vox* and *sermo*—sound, voice, and speech: animals "certainly have the capacity to make sound [*sonus*] and may even produce *vox*—sounds signifying to one another some interior state or *affectus* such as grief, joy, or fear—not all have the capacity to express concepts through articulate speech [*sermo*]."[36] Thus Blundeville's and Markham's assertion that the horse could learn to take meaning from a word through experience fits comfortably within Albert's theory: according to him, animals can be trained, and language can be used during training, but it is memory (in the limited sense of an animal's memory) and not understanding that is brought into play.[37] As Markham noted, the trainer should use "but one certaine worde of encouragement, for as the vse of many wordes . . . makes him he can neither vnderstand any word . . . so the vse of one single worde certainely, to one purpose, makes the Horse as perfitly by custome know the meaning thereof as your selfe that speakes it."[38] The horse learns through the immediate stimulus of the word, it has no access to abstraction, to imagined meanings; the word must always be used to always be useful. In this context Girolamo Cardano wrote, "dogges know theyre owne names, & are by teachinge learned to hunt fowles: but all these things they keepe by memorye, and trulye cannot tell how to put or chaūge any thing more then they be taught, what occasion soeuer they haue."[39]

But this is not the end of the debate about the nature of training a horse. Another idea also emerges in Markham's description of training methods: that is the more complex assertion of the animal's capacity to understand the training offered by the human. The horse, it would seem, can learn from the pain; it can compare past punishments with present actions and plan for a painless future. In fact, what becomes clear is that a horse might be capable of being prudent (comparing past and present to plan the future), and this possibility is brought to the fore in an alternative conception of the horse's intelligence.

In 1609 Nicholas Morgan wrote that "no lawfull and humaine Arte can effect any thing against nature."[40] These words do not come from a text dealing with magic; they come from a horse training manual, and Morgan is firm in his belief that training is natural and, more importantly, that it is an ex-

[36] Irven M. Resnick and Kenneth F. Kitchell Jr., "Albert the Great on the Language of Animals," *American Catholic Philosophical Quarterly* 70, no. 1 (1996): 42.

[37] Irven and Kitchell, "Albert the Great," 56.

[38] Markham, *Cavelarice*, VIII, p. 29.

[39] Girolamo Cardano, *Cardanus Comforte translated into English* (London: Thomas Marshe, 1573), sig. Ciii[v].

[40] Nicholas Morgan, *The Perfection of Horse-manship, drawne from Nature; Arte, and Practise* (London: Edward White, 1609), 6.

ercise of dominion. In fact, he argues that "man must consider that by his disobedience, he hath lost al obedience, which by original creation was subiect vnto him, & that now the obedience of all creatures must be attained by Arte, and this same preserued in vigor by vse and practise." To train a horse in these terms is to artfully return to a prelapsarian relation with animals. But one cannot work against the natural order; "knowledge and arte" (horse training) are "without nature, fondnes."[41] Thus, to perform this godly work it is necessary to interpret and choose the God-given tools correctly. To attempt to train a jade—what Blundeville was doing with the aid of a cat on a stick—is not only to waste one's time but also to be unnatural.

This emphasis on working *with,* rather than entering into a combative relation *against,* nature is most clearly to be traced in Morgan's fifth chapter, entitled "*An incytation to all men to loue Horsemen and Horses, and thereby the Arte.*" Here he calls for "simpathy of obedience & reciprocall loue" between horse and rider, "each accompaning the other in skill and ready attendance to the skil, as your eye and sence shall not apprehend, but one sence and one will by indiuiduall connexion of two bodies in one." This is a vision of art (the human) and nature (the horse) in perfect—prelapsarian—harmony, and this perfect harmony means that the kind of violent punishments laid down by Blundeville should not be needed. Morgan writes of "the man euermore commaunding the horse as his owne limmes" and states that "euery Horse is created as man is of soule and bodie, and is compounded of the foure elements as man." He goes on: "as man (beeing an vnderstanding creature) by often compulsarie abuses, may be made tame to be abused, so a creature void of reason and vnderstanding, may seeme to abide and indure that which nature abhorreth."[42] The torture of a horse does not distinguish human from equine, but instead reveals the similarity of rider and ridden. A horse may not have an immortal soul, but there is much that the two beings share; and it is the *shared* capacities once again—not difference or superiority—that enable humans to train a horse.

Morgan, then, like Blundeville, does not step beyond the orthodox ideas of the discourse of reason in his outlining of training methods, but he does reveal a rather different construction of the relation between human and animal. Here, rather than attempting to subdue nature—to beat resistance out of the horse—Morgan prefers to see a communication between horse and rider, and an enhancement of nature by art. Training a horse, Morgan argues, allows humans to concretize their status (to literally be on top of animals; to actually control the chariot) and to return to the prelapsarian relation in which animals were obedient and humans had absolute dominion.

---

[41] Ibid., 211, 5–6, 2, 56.
[42] Ibid., 15, 58, 63.

The Thomist argument offered a reason to treat animals with some degree of generosity without advocating anything like a kindness to animals for their own sake; Thomism, in fact, maintained human centrality. This is the argument that Morgan is following here: it is self-serving to be kind to a horse.

It is this sense of the horse as a being not so different from the human that is at work, I think, in the belief that the horse can be trained; in the unquestioned assertion that, in spite of their lack of reason, animals are capable of learning. The methods used in training a horse respond to the object status of animals, to animals' perceived natural (in)capacities—their wholly organic nature. But these methods also imply that animal nature is not quite as fixed as it is often represented as being; that change is not only for humans.[43] Morgan's context for this idea is the obvious one, and is one that makes such change a *return* rather than a development. For Morgan, horse training is a restoration of the original order of things when animals were obedient to humans. But this Christian context is only one way in which meaning can be given to the capacity of animals to be trained. Animals' capacities to experience and to move within the world (both capacities of the sensitive soul), in fact, can, like King James's dog, have more of intelligence in them than the discourse of reason might appear to allow: understanding is a capacity that is, in dangerous ways, shared by both man and beast. Even if a verbal conversation between Bankes and Morocco can only take place in the fabled context of the pamphlet, a communication between human and animal is vital in the training of that animal; in, paradoxically, the exercise of dominion over that lower creature.

At one point in his horse-training manual Morgan seems to make this dangerous conjunction of equine and human clear when he writes, "when the Ryder beginneth to teach a young horse, or a young Scholler, let him follow the order of a discreete Schoole-maister, that teacheth Children to write."[44] Here is a slippage from horse to child that Morgan seems to be aware of, even as he is oblivious to its connotations. It would be easy to say that Morgan is merely using a figure of speech and is implying no actual connection; one could claim that the distinction between child and horse has always already been made on the basis of the child's possession of an inorganic soul. It could also, of course, be argued that at the start of its education the child was like an animal and was then trained to be a human. But some discussions of education from the early modern period would seem to

---

[43] This is also the case in manuals for the training of hounds and hawks. See, for example, Estienne, *Maison Rustique;* George Turbervile, *The Noble Arte of Venerie or Hvnting* (London: Henry Bynneman, 1575), and *The Booke of Faulconrie or Hauking, For the Onely Delight and pleasure of all Noblemen and Gentlemen* (London: Christopher Barker, 1575).

[44] Morgan, *Perfection of Horse-manship,* 173.

upset such simple solutions. Owen Feltham, for example, complains that the man who places too much reliance upon others' ideas—who, in short, engages in too much book learning—"*argues* a barrennesse in *himselfe*, which forces him, to be ever a *borrowing*." Feltham makes a distinction between "*Iudgement*" and "*Reading*," between abstract (inorganic) and simple (organic) learning.[45] Memorization is not the same as understanding; a horse can do the former, and the emphasis on rote learning in the school system and the Church, along with the authority that remained with classical writings that were often the source of ideas in this period, means that the link between child and horse made by Morgan may have more perilous connotations than an accusation that I am simply overinterpreting his phrase might imply.

The early education of children and the education of horses were potentially not wholly different, and it thus becomes clear that the claim for the importance of the human's inorganic capacities of judgment, intellect, and will can be overstated. Animals, so this tradition states, do not have these capacities, but the human can only give evidence of expressing these capacities in actions like the exercise of judgment; at the same time, by implication, a failure to exercise judgment reveals the human to be dangerously animal. Rote learning gives no opportunity for judgment. As John Dod and Robert Cleaver noted: "let [the schoolboy] have the words taught him when he is able to heare and speake words, and after, when he is of more discretion, he will conceive & remember the sense too."[46] Parroting precedes speaking, and a parrot, of course, is not a human. Both training a horse and training a human can involve the immediate and refuse the abstract, with subsequent actions being the only means of judging difference. The problem that Morocco (and presumably any other horse trained using Markham's methods) presents is that in his actions he *appears* to judge, *appears* to exercise reason; he seems, after all, to look beneath the surface and tell a virgin from a whore. And it is the *appearance* of these capacities that creates such difficulties for the separation of human and animal.

The implications of this possibility, and the discursive context in which it was placed, are outlined by John Taylor, the Water Poet. In his "description of a Poet and Poesie, with an Apollogie in defence of Naturall English Poetrie," he writes,

> By teaching Parrots prate and prattle can,
> And taught an Ape will imitate a man:

[45] Owen Feltham, "Of Wisdome and Science," in *Resolves*, 130–31.
[46] John Dod and Robert Cleaver, *A Treatise or Exposition Upon the Ten Commandments* (London: Richard Bradocke, 1603), sig. 8ʳ.

> And *Banks* his horse shew'd tricks, taught with much lab[or]
> So did the hare that plaid vpon the tabor.
> Shall man, I pray, so witlesse, be besotted?
> Shall men (like beasts) no wisdome be allotted,
> (Without great studie) with instinct of Nature,
> Why then were man the worst and basest creature?

This is the danger of one way of viewing the apparently reasonable animals of early modern England. If, through education, animals can appear to perform human actions, and if humans require a similar education to be wise, then human superiority has vanished. Taylor, while recognizing this as a possible and logical outcome of the humanlike actions of animals, resorts, almost inevitably, to an assertion of difference that no amount of education can ever overturn:

> But men are made the other creatures Kings,
> Because superiour wisdome from them springs.[47]

In the context of discussions of Morocco, however, this assertion of the absolute superiority of humanity begins to come undone. It is an assertion that seems to be at odds with the evidence before it.

The works of Blundeville, Markham, and Morgan on training, then, might seem to offer a solution to the existence of the intelligent horse, but in attempting to naturalize what appears dangerously unnatural, training manuals reveal not only (by analogy) the limitation of the education of children, but also the apparent equivalence of human and animal scholars. Two attempted solutions, then—the unnatural and the natural; the magic and the pedagogic—offer answers to the conundrum of animal reason but also reveal frailties in the condition of those attempting the solution. In neither case is the "intelligent" animal comfortably explained. Yet another issue of some importance takes us into a very different realm, both intellectually and socially. The very popularity of Morocco reveals a fascination, I think, with the mere potential for animal intelligence in the period, a fascination that is itself worth taking seriously. In fact, it is in the public interest, as much as in the philosophical, theological, demonological, and pedagogical writings of the period that the import and threat of animal reason can be traced. The intelligent animal was something that concerned more than merely the educated elite; this was an issue that was being considered at all levels of society.

---

[47] John Taylor, *The Nipping and Snipping of Abuses* (London: Ed. Griffin, 1614), sig. B4[r].

Knowing Knowledge

In 1614 Samuel Rid repeated Reginald Scot when he revealed how "to make a horse tell you how much money you haue in your purse." Rid lifts the story of "an asse at *Memphis* in Egypt, that could doe rare feates" almost verbatim from Scot.[48] This ass could play dead, perform on cue the emotions of joy and sadness, and—what for Rid is almost the most important thing—earn his owner a decent sum of money. "Such a one is at this day to be seene in London," he writes, and once again Morocco appears as a case worthy of examination.[49] Rid describes the horse's counting abilities—what Joseph Hall called "strange Morocco's dumb arithmetick"[50]—and other tricks, and comments that "not one among a thousand perceiues how they are done, nor how he is brought to learne the same." This is where the success of the act lies: if the performance was transparent, if it was clear that it was not a display of true reason, then the act would not be such a money-spinner. A thought that takes effort—that is drawn out of the thinker by persuasion—is not a thought at all; it is probably a guess, and the difference is an important one. If you recall, Scot argued that accusations against witches were not true but blasphemous; what he called "*ghesses, presumptions, & impossibilities contrarie to reason, scripture and nature.*" The structure of Scot's phrase creates a link between a guess and its opposite, reason—the rationale being that reason involves judgment, and a judgment is, if not antithetical to, then at the least very different from, a guess. As Nemesius wrote, "The workes of *chance* are such as befall *unreasonable,* and *inanimate* creatures, without *nature,* or *art.*"[51]

But, of course, Samuel Rid's intention is, he writes, to "rip vp some proper trickes," and to reveal the sleight of hand behind some of the most impressive miracles of the age. Like the other side of the coin to Scot and Markham, he wants to reveal not how a supernatural thing is actually wholly natural but how "a naturall thing be made to seeme supernaturall," and

---

[48] S. R. [Samuel Rid], *The Art of Jugling or Legerdemaine* (London: G. Eld, 1614), sig. F4ᵛ. The story is told in Scot (*The discoverie*, 253–54) as another attack on Bodin's beliefs in actual transformation. This story of the ass, like Scot's tale referred to earlier, has a significant textual history. Rid (via Scot) tells of the ass's performance: it faked death when the threat of hard labor was voiced, but its counterfeit end was revealed when a day of celebration was announced and it greedily leapt up. In a similar vein Plutarch tells the story of a dog in Rome who was apparently poisoned and then came back to life. Plutarch, *Whether Creatvres Be More Wise, They of the Land, or Those of the Water,* in *The Philosophie, commonlie called, The Morals Written by the learned Philosopher Plutarch of Chærnea* (London: Arnold Hatfield, 1603), 967.

[49] [Rid], *Art of Jugling,* sig. F5ʳ.

[50] Joseph Hall, *Virgidemiarum: Satires in Six Books* (1597; repr., Oxford: R. Clements, 1753), Book IV, Satire II, p. 62.

[51] Nemesius, *The Natvre of Man* (London: Henry Taunton, 1636), 542.

Rid's unmasking of the intelligent horse is simple in the extreme: "note," he writes, "that nothing can be done, but his master must first know, and then his master knowing, the horse is ruled by him by signes. This if you marke at any time you shall plainely perceiue."[52] Or, as Thomas Sebeok has noted more recently, ask the horse a question to which his owner does not know the answer, and you will soon discover whether the animal is intelligent or not.[53] This is, of course, remarkable in its simplicity, and the fact that this simple explanation is only fully developed by Rid, a celebrator of legerdemain, opens up the possibility that in Morocco, this popular—even notorious—figure from the entertainment scene in early modern England, we can begin to piece together some alternative ways in which people understood the rational capacity of animals.

You only ever see what you expect to see. This is a point that, again, comes from Sebeok, but reflects the conception of animals found, I have suggested, in a number of purportedly factual representations of them in the early modern period; perhaps Morocco also mirrors what is taken to be true by his human audience. The game played by Bankes and his spectators is based on mutual agreement: he knew that they knew that this was a pretended intelligence—a "guess"—but he also knew that they were willing to accept the possibility of animal intelligence. Moving beyond the paradigm of the beast fable, where the animal had a voice in order to represent a single human quality, Morocco had only "*dumb* arithmetick." But it was the arithmetic, not the lack of voice, which was so impressive, because the performance of intelligence—the stamp of the foot—is itself a kind of language, is itself an expression of reason. When Morocco counts the spots on the dice, adds up the coins in someone's hand, or picks the virgin from the whore, he is enacting his intelligence, displaying it, and the audience does not have to assume the existence of a thought—it is there before them: it is what they have paid to see. If language, as Ralph Lever proposed, is merely an expression of the reasonable capacity of the speaker,[54] then Morocco's audience is able to believe that he has such a capacity. "Strange Morocco's dumb arithmetick" is strange in that it is wonderful—an apparent miracle of nature—but it is also strange because it appears to be true.

This mingling of knowing and guessing when watching the performance of the intelligent horse can be observed in a contemporary anecdote. *Tarlton's Jests* is a collection of tales about the life of the comedian Richard Tarl-

[52] [Rid], *Art of Jugling*, sigs. F5[r], B3[r], B2[v], and G1[v].

[53] Thomas A. Sebeok, *The Sign and Its Masters* (Austin and London: University of Texas Press, 1979), 91. I owe this reference to John Simons.

[54] Ralph Lever, *The Arte of Reason, rightly termed, Witcraft, teaching a perfect way to argue and dispute* (London: Henrie Bynneman, 1573), sig. *v[r].

ton, collected posthumously in 1638. And in this text a story is told of Tarlton's meeting with Bankes' "Horse of strange qualities" at the Cross Keys Inn, London. Bankes spots Tarlton in his audience, and:

> (to maketh the people laugh) saies *Signior* (to his horse) Go fetch me the veryest foole in the company. The Jade comes immediately, and with his mouth drawes *Tarlton* forth: *Tarlton* (with merry words) said nothing, but *God a mercy Horse*. In the end *Tarlton* seeing the people laugh so, was angry inwardly, and said, Sir, had I power of your horse, as you haue, I would doe more then that. What e're it be, said *Bankes* (to please him) I wil charge him to do it. Then (saies *Tarlton*) charge him to bring me the veryest whore-master in this company. He shall (saies *Banks*) *Signior* (saies he) bring Master *Tarlton* here the veryest whore-master in the company. The horse leades his Master to him. Then God a mercy horse indeed, sayes *Tarlton*.[55]

Picking out Tarlton is an expected part of the act: distinguishing Englishman from Spaniard, virgin from whore, was a skill for which Morocco was well known. Markham wrote that in order to achieve this the trainer must point his rod toward the chosen person, and "to that place also you must most constantly place your eie, not remoouing it to anie other obiect, til your wil be performd, for it is your eie and countenance, as wel as your words, by which the horse is guided."[56] Samuel Rid likewise noted that "the eye of the horse is alwaies vpon his master."[57] Similarly, Thomas Dekker wrote "(as *Bankes* his horse did his *tricks*) onely by the eye, and the eare."[58] There is no recognition or judgment made by the horse; there is merely the work of the senses and a learned following of orders. These writers have undermined the possibility of animal reason: here there are immediate stimuli—the rod, the eye—to which the animal merely responds.

But Morocco's selection of Bankes is rather different: "had I power of your horse, as you haue," Tarlton says; the implication being that the actions all come from the human not the animal—which is Markham and Rid's line. But when the horse names his owner a "whore-master," he undercuts the dominion which Bankes' control is meant to reinforce. When he presents his master as a pimp, Morocco is uprooting all order, and for a second it appears that the horse has answered a question to which his owner did not know the answer. For a second. You see what you expect to see.

---

[55] Richard Tarlton, *Tarltons Jests* (London: Ed. Griffin, 1638), sigs. C2r–v.

[56] Markham, *Cavelarice*, VIII, p. 32.

[57] [Rid], *Art of Jugling*, sig. G1v.

[58] Thomas Dekker, *The Seuen deadly Sinnes of London* (London: Nathaniel Butter, 1606), 35.

The people had much ado to keepe peace: but *Bankes* and *Tarlton* had like to have squar'd, and the horse by to giue aime, But ever after, it was a by-word thorow *London, God a mercy Horse,* and is so to this day.[59]

Rid argued that "a Jugler must set a good face vppon that matter he goeth about, for a good grace and carriage is very requisite to make the art more authenticall."[60] What could more authentically prove the power of a horse than for the horse to turn against the man who is meant to control him? By naming himself a "whore-master" Bankes shows good grace—his modesty permits him to acknowledge that his job, displaying the intelligent horse, is base in the extreme. But he also shows good business sense; he gets his horse to admit his whorishness, and in that admission gets more fame for his animal.

In Tarlton's "*God a mercy Horse,*" however, we might also see something akin to the skeptical assertion of tranquility. Tarlton, of course, does not continue to inquire—does not ask, "how did you do that?"—rather, he acts as if he accepted the possibility both that the horse's judgment was the *horse's* judgment and that Bankes had power over his animal. He reaches, you might argue, a state of undecidability and retires to a tranquil acknowledgment that he can never truly know. This reaction is the one the audience wants because it is part of their "contract" with Bankes and Morocco. Watching a man work with a well-trained horse is not unusual—you might see it every day walking around the city—but watching a man work with what appears to be a horse with a reasonable mind of its own is quite another matter. And it is the latter that the audience has paid to see. Tarlton, the great performer, knows this only too well.

But perhaps, while proposing that Tarlton's "*God a mercy horse*" is a staged performance of skepticism, it is also worth thinking about why Morocco was so popular; why a horse that could count coins could earn his owner a decent living and could inspire so many references in contemporary writing. Perhaps the audiences who gathered to watch this trained animal were themselves engaging in a kind of skepticism. They would wonder about the counting horse and would perhaps invoke other animals—the birds that built nests; the spiders with their webs; the dogs that could round up sheep, hunt hares, and so on—to make sense of him. But the sense that might come from these wonderings might challenge established ideas, might work against the belief in the unreasonable nature of animals, and might shift the focus from ideas to reality, from philosophizing to living. The animals that could be used to explain Morocco existed in the world outside of books, out-

---

59 Tarlton, *Tarltons Jests,* sig. C2ᵛ.
60 [Rid], *Art of Jugling,* sig. B3ᵛ.

side of intellectual discussions. They could be found in a world available to all, and meaningful to all. In this context—in distinction from the magical or pedagogical explanations for Morocco—what Morocco was capable of was not so surprising. Anyone who owned a horse would know the animal's capacities; anyone who had a dog would likewise know that it was not simply irrational, but was useful and capable of displaying what looked like intelligence—was a colleague as much as, if not more than, an object. As the author of *The Sceptick* wrote: "The dog . . . we see is plentifully furnished with inward discourse."[61]

To state that animals are inferior to humans is to state something that, within a Christian, Aristotelian framework, appears to require no further explanation: the existence of dominion and the possession of the rational soul make such unnecessary. However, what is clear in numerous texts available in early modern England is that such a statement of animal inferiority is not always present—not only because of the breakdown in the logic of the discourse itself or the reemergence of a lost philosophical tradition, but because all the complications of humans' everyday existences alongside animals challenge it. In this sense Morocco comes to represent what was already known. Animals are not the same as humans, but that does not mean that they are incapable, or that they are not in themselves subjects. When Morocco stamped his feet and picked out his owner as a "whore-master," he was merely performing in public the kinds of animal capacities that people were aware of in their personal experiences with animals. Perhaps these experiences are so mundane that they are rarely recorded or discussed. Perhaps the orthodox philosophical debate sits at odds with what was apparently obvious in day-to-day living. Animals think.

I am not saying that Morocco was an intelligent horse, or rather I am not arguing that he could count, understand the English language, and tell a virgin from a whore. Nor am I arguing that early modern audiences were incapable of telling the difference between a trained horse and a truly intelligent one. What I am saying, however, is that the constant references to Morocco in critical and comic writings from the period, and his popularity in performance, show that in the early modern period the possibility of an animal with the capacity to reason was present in some important ways. Far from being a time that either celebrated or simply assumed animal irrationality, it was a period clearly aware of a problem within the discourse of reason. The simple distinction between reason and unreason that so many thinkers proclaimed was constantly being eroded in a variety of ways, and what can be traced in its place, perhaps, is a belief in the power of Morocco to turn against and to truly name his master. Such a possibility can be found

[61] *The Sceptick*, 47.

not only in the writings surrounding Morocco but also in King James's assertion that "a hound had more in him than was imagined"; in Wright's capable lamb; in Goodman's vision of the pleasure-seeking hawk.

These visions of animals, however, coming as they did from very different traditions—skeptical, empirical, Plutarchian, Montaignean—were all silenced by one man in the 1630s, and the means by which he silenced these ideas, and the ways in which they returned and persisted, form the focus of the next chapter.

CHAPTER 6

A Reasonable Human?

In a work that was first published in French in June 1637, translated into Latin in 1644 and into English in 1649, the French philosopher René Descartes described his attempts to arrive at a new science. *Discourse on the Method* takes his readers through the rules of his new system, its moral implications, his evidence for the existence of God and the human (reasonable) soul, and, in part 5, the nature of animals. Descartes's "beast-machine" hypothesis, first outlined in print here, offers a new vision of the difference between humans and animals.[1] It is worth quoting from Descartes's text at length to give a sense of the novelty of his ideas.

> I made special efforts to show that if any machines had the organs and outward shape of a monkey or of some other animal that lacks reason, we should have no means of knowing that they did not possess entirely the same nature as these animals; whereas if any such machines bore a resemblance to our bodies and imitated our actions as closely as possible for all practical purposes, we should still have two very certain means of recognizing that they were not real men. The first is that they could never use words, or put together other signs, as we do in order to declare our thoughts to others. . . . Secondly, even though such machines might do some things as well as we do them, or perhaps even better, they would inevitably fail in others, which

---

[1] A similar theory in which animals were compared with clocks had been proposed by the sixteenth-century Spanish philosopher, Gomez Pereira. Descartes denied reading his work. See Stephen Gaukroger, *Descartes: An Intellectual Biography* (Oxford: Oxford University Press, 1995), 271.

would reveal that they were acting not through understanding but only from the disposition of their organs.[2]

For numerous early modern thinkers, as the various discussions of Morocco revealed, the question of the existence and nature of animal rationality was either (at its foundation, anyway) simple: animals were unreasonable, humans reasonable. Or the question was unanswerable because of the writers' skepticism, because of their lived relationships with animals. Descartes's statement of animal automatism, however, seemed to present a final solution. Shifting away from Aristotelianism, Plutarchianism, and skepticism, Descartes offered his mid-seventeenth-century readers a vision of animals in which they were no longer unreasonable, reasonable, or even proto-reasonable; Descartes's animals were automata. It is not intelligence, he wrote, but "nature that acts in [animals] according to the disposition of their organs. In the same way a clock, consisting only of wheels and springs, can count the hours and measure time more accurately than we can with all our wisdom."[3]

In this chapter, I want to read Descartes's beast-machine hypothesis in the light of the problems that faced philosophers writing within the discourse of reason. I also want to outline the arguments for Cartesian philosophy that were offered by a number of writers in the latter half of the seventeenth century, and to show how far the arguments for Cartesianism were founded in part on a perceived *need* to clarify the difference between humans and animals. What will come to the fore once again in this chapter are questions about the status of animals but also about cruelty, ethics, and, inevitably, the nature of the human in the face of its eternal others.

The New Human

In the relation between humans and animals outlined in the discourse of reason, humans, possessing reason, left animals in their wake. Whatever other problems emerged from the discourse—the possibility that a natural, innate humanness did not exist, the breakdown of human superiority in the face of animals' lack of vice, and so on—its starting point was straightforward. While beasts and humans shared vegetative and sensitive souls, it was the possession of the inorganic, rational soul that set human apart from an-

[2] René Descartes, *Discourse on the Method* (1637), in *The Philosophical Writings of René Descartes*, ed. John Cottingham, Robert Stoothoff, and Dugald Murdoch (Cambridge: Cambridge University Press, 1985), 1:139–40.
[3] Ibid., 141.

imal. For Descartes, the distinction was rather different; as he stated in a letter to Regius (the Dutch physician, Henri le Roy) in May 1641, "the term 'soul' is ambiguous as used of animals and of human beings."[4] What Descartes meant by this was that the movements of the (human or animal) body could be better and more clearly explained without recourse to the notion of vegetative and sensitive souls; and that the only soul present was the rational one, which allowed humans alone access to three capacities. The first was, inevitably, reason—using conventional terms, Descartes listed the faculties of reason as wit, imagination, and memory.[5] The second capacity was language, by which humans expressed their reason. And the third, volition, that is, the possibility of acting and reacting reasonably in infinite circumstances. Animals, like clocks, had skills, but those skills were limited to certain organic predispositions—a hound could scent, a bird could migrate, a cat could catch mice, and so on. Humans, however, while they might lack the olfactory talents of a dog, the natural navigational skills of birds, and the cat's capacity to catch vermin without recourse to extrabodily instruments, were not limited as animals were by their bodies. Humans also had will, the capacity to respond to situations and to choose those responses. But the human was, of course, not only a willing being; humans shared with animals the body, and thus, as Descartes wrote in his last work, *The Passions of the Soul:*

> every movement we make without any contribution from our will—as often happens when we breathe, walk, eat and, indeed, when we perform any action which is common to us and the beasts—depends solely on the arrangement of our limbs and on the route which the spirits, produced by the heat of the heart, follow naturally in the brain, nerves and muscles. This occurs in the same way as the movement of a watch is produced merely by the strength of its spring and its configuration of wheels.[6]

This emphasis on spirits in the brain, nerves, and muscles appears to echo in some ways the belief that humans and animals share in the capacities of the vegetative and sensitive souls. However, for Descartes the absence of reference to these souls was not merely a change in terminology; rather, he places emphasis on the body itself (rather than the soul) as the source of unthought actions.

Self-consciousness, so important in the discourse of reason, remained central for Descartes, but his is a notion of self-consciousness that is rather different from that of the man who knows he is a man and is therefore dif-

---

[4] Descartes to Regius, May 1641, in *Philosophical Writings*, 3:181.
[5] Descartes, *Discourse on the Method*, 111–12.
[6] René Descartes, *The Passions of the Soul* (1649), in *Philosophical Writings*, 1:335.

ferent from the horse who has no conception of its horse-ness. For Descartes, self-consciousness was the beginning of science. In *Discourse on the Method* he describes how he set out to find a firm footing for this intellectual inquiry, and while beginning with skepticism, he ended in a way that undermined the skeptical emphasis on the unknowability of anything.

"What do I know?" was the key question for Montaigne. In answer, he found that he knew—absolutely knew—very little. And this lack in him led to his distrust of humans and their structures of thought. Descartes, on the other hand, while using skeptical foundations for his theory, was careful to offer something that no true skeptic could ever offer: a conclusion. He began by doubting everything, but wrote of his own doubting, "In doing this I was not copying the sceptics, who doubt only for the sake of doubting and pretend to be always undecided; on the contrary, my whole aim was to reach certainty—to cast aside the loose earth and sand so as to come upon rock or clay."[7] Beginning with absolute (Cartesian) doubt, then, and questioning the reality of even his own body, the certainty Descartes reached was "that I, who was thinking [that everything was false], was something. And observing that this truth, '*I am thinking, therefore I exist*' was so firm and sure . . . I decided that I could accept it without scruple as the first principle of the philosophy I was seeking."

But, of course, in his regimen of doubt Descartes was able to cast aside many of the ideas that were accepted by other thinkers of his age. Perhaps most crucially, the close connection between the body and the soul, so central to discussions about the immorality of ill health, for example, was severed by Descartes:

> I knew I was a substance whose whole essence or nature is simply to think, and which does not require any place, or depend on any material thing, in order to exist. Accordingly this "I"—that is, the soul by which I am what I am—is entirely distinct from the body, and indeed is easier to know than the body, and would not fail to be whatever it is, even if the body did not exist.[8]

This dualism offered a solution to the precarious status of the human of the discourse of reason, who was a being locked in a constant struggle with his or her physical being; whose will could be overrun by passions; who could become a beast. Instead, for Descartes, the self was to be found only in the mind, and the body was merely a temporary wrapping for this true self, on which the body had no significant impact. In this context descent to the beast, so fearful a presence in the discourse of reason, was not only implau-

---

[7] Descartes, *Discourse on the Method*, 125.
[8] Ibid., 127.

sible—it was impossible, since the human was always human. In fact, in opposition to the logic that had emerged from the Aristotelian model—according to which logic some humans were not actually fully human—Descartes argued that full human status was already possessed by all humans. In the opening paragraph of *Discourse on the Method* he wrote:

> the power of judging well and of distinguishing the true from the false—which is what we properly call "good sense" or "reason"—is naturally equal in all men, and consequently . . . the diversity of our opinions does not arise because some of us are more reasonable than others but solely because we direct our thoughts along different paths and do not attend to the same things. For it is not enough to have a good mind; the main thing is to apply it well.[9]

While in the (earlier) discourse of reason the starting point may well also have been that all humans have reason, what had swiftly emerged was the fact that the *display* of reason was imperative. In that framework there was a slippage from a simple assertion of the a priori possession of the rational soul to the necessity of empirical evidence of its existence in the individual. The reason displayed as proof of possession of the rational soul displaced the argument for the innate possession of that soul, and so reason was represented as something that was learned, was evidenced in virtuous actions, and could be lost through vicious behavior. In Descartes's writings, however, that slow and somehow-not-quite-inevitable progress toward the true human status of the "good mind" found in the discourse of reason was gone, and in its place was his unwavering assertion that all humans possess reason. This belief provides a very different basis on which to build a philosophy; a basis that has implications for understanding what he writes about animals.

No More Empiricism

Katherine Morris has outlined a core conception of this philosophy: "within Descartes' framework," she writes, "possession or non-possession of a rational soul is part of something's *nature* or *essence*. The nature of a creature is not something to be empirically determined; it is at two removes from its observable 'behaviour.'" In the Cartesian view, to see a dog act reasonably, as James VI and I argued that he had, did not in fact challenge the structures of thought that presented animals as irrational. Rather, it was the preexisting nature of the dog—which for Descartes lacked reason—that was central

[9] Ibid., 111.

to comprehending it. The emphasis placed by the orthodox discourse of reason on seeing reason in action is gone. Morris writes that a machine that "behave[s] in as complex a way as human beings do" remains possible within Descartes's schema and yet does not compromise his notion of the human because "*given* that it is a machine . . . it *could not* be exhibiting the power of thinking."

Looking at animals' behavior, then, as a way of understanding the rationality of those animals is irrelevant. Animals are animals, and humans are always already humans, and the Plutarchian emphasis on nest building, cunning, the natural possession of medical knowledge, and so on, in no way challenges the status of humans or of animals. The writings of Thomas Rogers, John Maplet, and Godfrey Goodman are cast aside, and what emerges in their place is an assertion of absolute difference, of an unbridgeable, natural chasm between humans and animals.

Reiterating Descartes's emphasis on preexisting status, Morris writes that "natures (hence faculties) belong to species, not to individuals." The implications of this statement are twofold:

> on the one hand, if it is *known* that the individual belongs to a species whose nature does not include possession of a rational soul, then no matter how apparently intelligently the individual responds to the contingencies of life . . . the individual is not exhibiting intelligent behaviour. . . . And on the other hand, if it is *known* that an individual *does* belong to a species which by nature has a rational soul, then no matter how apparently witless its performance . . . the individual does have the faculty of thinking and hence possesses a rational soul.[10]

In this context the child is always already human, even though it cannot yet display its possession of reason; and the beastly human disappears from view. The challenges that the discourse of reason presented to itself are gone. In his reply to Pierre Gassendi's objections to his *Meditations*, for example, Descartes writes that men who followed someone else without recognizing that the person they followed possessed superior judgment "would be behaving more like automatons or beasts than men."[11] Here, the thoughtless action—which had, in the discourse of reason, logically undone the status of the human—is merely a failure to apply reason. But because the human is always already human, such a failure does not disrupt status, does not

---

[10] Katherine Morris, "*Bête-machines,*" in *Descartes' Natural Philosophy,* ed. Stephen Gaukroger, John Schuster, and John Sutton (London: Routledge, 2000), 407–8.

[11] René Descartes, "Appendix to the Fifth Objections and Replies," in *Objections and Replies,* in *Philosophical Writings,* 2:273.

undo one's humanity. These foolish humans behave "*more like*" beasts, they do not become beasts.

By proposing a metaphysical foundation from which to build a new science, Descartes was able, then, to formulate an absolute and, within the context of his discourse itself, unquestionable distinction between humans and animals. This is something that the discourse of reason had failed to do. In fact, criticism of the beast-machine hypothesis that came from outside of the metaphysical foundations of Descartes's theory could straightforwardly be dismissed by that theory. So what if a dog looked like it was performing a syllogism? Experience, custom—whether from nature or training—could account for such a phenomenon; no reason was implied because no reason was ever present.

But the error of King James was also a common one. The experience of seeing an animal offered dangerously plausible evidence for animal rationality. As Descartes himself wrote in a letter to the French philosopher Reneri (Henri Regnier) in April/May 1638,

> Most of the actions of animals resemble ours, and throughout our lives this has given us many occasions to judge that they act by an interior principle like the one within ourselves, that is to say, by means of a soul which has feelings and passions like ours. All of us are deeply imbued with this opinion by nature.[12]

Likewise Sir Kenelm Digby, author of the first Cartesian study, *Two Treatises* (1644; to which I return in my conclusion), wrote at the end of the final chapter of *A Treatise of Bodies,*

> this is a generall and maine errour, running through all the conceptions of mankind, unlesse great heed be taken to prevent it, that what subject soever they speculate upon, whether it be of substances, that have a superiour nature to theirs, or whether it be of creatures inferiour to them, they are stil apt to bring them to their own standard, and to frame such conceptions of them, as they would do of themselves.[13]

For the skeptic, this framing of the world as the self—also known as anthropomorphism—would signal the end of the human capacity to truly know animals (or indeed anything). For the Cartesian, however, the almost instinctive anthropomorphic thought processes of humans could be coun-

---

[12] Descartes to Reneri for Pollot, April or May 1638, in *Philosophical Writings,* 3:99

[13] Sir Kenelm Digby, *A Treatise of Bodies,* in *Two Treatises* (1644; repr., London: John Williams, 1645), 419.

tered by an act of will. Refusing to anthropomorphize, refusing to believe one's childish first impressions, was therefore crucial to a full understanding of the world and of Descartes's philosophy. As Descartes had written: "it is not enough to have a good mind; the main thing is to apply it well." Such an application, it would seem, *requires* the absence of anthropomorphism, *requires* that animals be automata.

But the Cartesian denial of reason to animals, a denial that went further than anything that was offered by the discourse of reason, did not merely establish a theoretical framework by which the behavior of animals could be (pre)judged. Descartes's ideas also offered comfort to humans by countering not only the failings of the discourse of reason but also some of the speculations of skeptical writings. Descartes, like the skeptics, doubted; but he argued that to doubt was to be, and thus he was able to formulate a conclusion rather than abandon the search in favor of tranquility. This was very different from the suggestion that animals were unknowable, that they were as likely as not to possess reason as other humans were.

Painless Separation

Language offered Descartes evidence of the lack of reason of animals, or rather an animal's lack of language merely proved what was already known: that the animal lacked reason. As Descartes wrote in a letter to William Cavendish in November 1646, "Montaigne and Charron may have said that there is a greater difference between one human being and another than between a human being and an animal; yet there has never been known an animal so perfect as to use a sign to make other animals understand something which bore no relation to its passions."[14] But lacking reason was not merely a question of silence; it was also a question of having, or not having, the status of a moral patient. Whereas Montaigne and some of those writing after him had regarded animals as fellow beings, perhaps the most significant ethical implication of Descartes's theory was his argument that animals could not experience pain in the way that humans could. In fact, in Descartes's view an animal's response to a painful stimulus was merely a mechanical one and did not reflect the kind of rational reaction shown by humans.

Descartes's beast-machine hypothesis and its ethical implications are famously clarified in the exchange of letters between Descartes and the English Neo-Platonist Henry More in 1648–49. More, who had earlier been

[14] René Descartes to the Marquess of Newcastle, 23 November 1646, in *Philosophical Writings*, 3:302 and 303.

influenced by Descartes's ideas, wrote to the French philosopher that "my spirit, through sensitivity and tenderness, turns not with abhorrence from any of your opinions so much as from that deadly and murderous sentiment which you professed in your *Method*, whereby you snatch away, or rather withhold, life and sense from all animals, for you would never concede that they really live."[15] Descartes's response to More's accusation was clear and reiterated something that he had already argued on numerous occasions: "there is," he wrote, "no preconceived opinion to which we are all more accustomed from our earliest years than the belief that dumb animals think." His explanation for this deeply held human belief is that "we see that many of the organs of animals are not very different from ours in shape and movements." This in turn, he argues, leads us to believe that the motivating force found in humans—reason—is also to be found in animals. However, this is wrong: the movements of animals, he writes, "could all originate from the corporeal and mechanical principle" and thus "we cannot at all prove the presence of a thinking soul in animals." Descartes acknowledges that this lack of proof of presence is not itself proof of absence: "though I regard it as established that we cannot prove there is any thought in animals, I do not think it can be proved that there is none, since the human mind does not reach into their hearts."[16] This, of course, echoes the argument of the skeptics: as the English text *The Sceptick* said: "I may tell what the outward object seemeth to me, but what it seemeth to other creatures, or whether it be indeed that which it seemeth to me or to any other of them, I know not."[17] However, the apparent lack of proof of animal irrationality that would lead any good skeptic to abandon the search and thus achieve tranquility had the opposite effect on Descartes. The only reason we believe that animals have thought, he writes, is that for us thought is "included in our mode of sensation" and therefore, because "it seems likely that [animals] have sensation like us," it is believed that they have thoughts like us as well. Such an argument, "which is very obvious, has taken possession of the minds of all men from their earliest age." But, Descartes continues, "there are other arguments, stronger and more numerous, but not so obvious to everyone, which strongly urge" that animals are irrational. "One is that it is more probable that worms, flies, caterpillars and other animals move like machines than that they all have immortal souls."[18]

[15] Henry More to Descartes, 11 December 1648, in Leonora D. Cohen, "Descartes and Henry More on the Beast-Machine—A Translation of their Correspondence Pertaining to Animal Automatism," *Annals of Science* 1 (1936): 50.

[16] René Descartes to Henry More, 5 February 1649, in *Philosophical Writings*, 3:365.

[17] William M. Hamlin, "A Lost Translation Found? An Edition of *The Sceptick* (c. 1590) Based on Extant Manuscripts [with text]," *English Literary Renaissance* 12, no. 2 (2001): 47.

[18] Descartes to More, 5 February 1649, 365–66.

It is significant, surely, at this point in his argument—which is based not on absolute, a priori evidence, as we might assume, but on probability and lack of proof—that Descartes turns to insects, to creatures who were, in Lodowick Bryskett's terms, "the most imperfect among liuing creatures."[19] Whereas, in *Discourse on the Method* he had argued that a machine with the outward shape of a *monkey* would be impossible to differentiate from a real monkey, here creatures whose status and ensoulment had always troubled philosophers are offered in evidence.[20] Descartes had already used this method of reductio ad absurdam in his letter to Cavendish, where he stated that if animals had "thought as we do, they would have an immortal soul like us. This is unlikely, because there is no reason to believe it of some animals without believing it of all, and many of them such as oysters and sponges are too imperfect for this to be credible."[21] Monkeys and sponges—these creatures are not differentiated by Descartes, whereas humans (all humans) stand above them all. This is, perhaps, one of the most vivid illustrations of Jacques Derrida's argument that it is the general singular concept, "the animal," that lies at the heart of "interpretive decisions."[22] If Descartes had distinguished between the oyster and the dog or the sponge and the horse, perhaps the credibility of arguments for the reasoning capacities of animals would not have been quite so easy to dismiss.

In his letter to More, however, Descartes moves from the statement that it is "more probable" that animals are machines than immortal beings to reiterate some points that he had made in *Discourse on the Method:* that animals and automata are indistinguishable; that animals lack "real speech." But it is at the end of the letter that we find Descartes's most direct response to More's assertion that the beast-machine hypothesis is "murderous" because it withholds "life and sense from all animals." Here Descartes returns to what I am taking as one of the key ethical claims of his theory. He writes:

> For brevity's sake I here omit the other reasons for denying thought to animals. Please note that I am speaking of thought, and not of life or sensation. I do not deny life to animals, since I regard it as consisting simply in the heat of the heart; and I do not even deny sensation, in so far as it depends on a bodily organ. Thus my opinion is not so much cruel to animals as indulgent

[19] Lodowick Bryskett, *A Discovrse of Civill Life* (London: Edward Blount, 1606), 190.

[20] Saint Augustine famously worried that the capacity of two halves of the same "many-footed" worm to live apparently independently when severed from each other would undo any direct relation between soul and body, and would thus destroy his argument, "because one little worm bored through it." St. Augustine, *De Quantitate Animae*, trans. Joseph M. Colleran (Westminster, Md., and London: Newman Press and Longmans, 1949), 90, 91–92.

[21] Descartes to the Marquess of Newcastle, 304.

[22] Jacques Derrida, "The Animal That Therefore I Am (More to Follow)," trans. David Wills, *Critical Inquiry* 28 (2002): 409.

to human beings—at least to those who are not given to the superstitions of Pythagoras—since it absolves them from the suspicion of crime when they eat or kill animals.[23]

Descartes seems to be arguing that animals can experience the world but that their experience is not a full one (that is, is not as complete as humans' experience). This incompleteness is because animals lack thought and, crucially, the self-awareness that allows humans to fully experience their experiences. In fact, Descartes's frequent use of the analogy of animals and clocks is not merely to be taken as a useful descriptive simile: animals, in Descartes's view, could live in the world, display certain natural—bodily—skills in that world, but could never actually experience that world in the way that humans could. A clock, after all, could tell the time better than a human, but it would fail miserably to, say, enjoy a good dinner in the company of others when that time arrived.

Thinkers writing within the discourse of reason would never deny that animals could experience the world. Even if that world was available to animals only in the present—that is, even if animals were believed to lack access to the past and the future without direct stimulation—that present world impacted on animals in a way that meant that an "experience" of that world could be ascribed to them. When in horse-training manuals, to offer an obvious example, whips were used as part of training, clearly the trainers believed that animals would experience the pain from the whip and respond in a way that revealed their desire not to experience that pain again. Descartes's theory, however, offered his readers something new. A clock does not "know" the time, it merely records it, and likewise, an animal did not "know" it lived, rather it merely lived.

The modern philosopher and translator of Descartes, John Cottingham, has succinctly summarized Descartes's view: "I should certainly say that cats feel pain, but not that they have the kind of full mental awareness of pain that is needed for it to count as a *cogitatio*."[24] That is, that an animal can experience torture, but cannot somehow properly *feel* it. Or, as Descartes himself put it in a letter to Marin Mersenne in June 1640: "pain exists only in the understanding," whereas in an animal's response to an apparently painful stimulus it is external movements alone "which occur, and not pain in the strict sense."[25] It is for this reason that the beast-machine hypothesis, as Descartes wrote to More, is "indulgent to human beings." Humans can inflict pain on animals without actually, well, inflicting pain.

[23] Descartes to More, 5 February 1649, 366.

[24] John Cottingham, "'A Brute to the Brutes?': Descartes' Treatment of Animals," *Philosophy* 53 (1978): 558.

[25] Descartes to Marin Mersenne, 11 June 1640, in *Philosophical Writings*, 3:148.

This aspect of Descartes's ideas—his argument that the animal's experience of pain was not a full experience of pain—was used to support the increase in vivisection in France. In his *Mémoires pour servir à l'histoire de Port-Royal* (1738), for example, Nicolas Fontaine now infamously recorded how at that revered institution

> They administered beatings to dogs with perfect indifference, and made fun of those who pitied the creatures as if they had felt pain. They said that the animals were clocks; that the cries they emitted when struck, were only the noise of a little spring which had been touched, but that the whole body was without feeling. They nailed poor animals up on boards by their four paws to vivisect them and see the circulation of the blood which was a great subject of conversation.[26]

Such activities and the "indifference" of those involved in them were, as Descartes had proposed in his letter to Henry More (even if he had not intended this use of his hypothesis), to be "indulged" because animals were automata.

Descartes's idea was not, however, taken up by all, even though it "excused" human activities such as vivisection. There were many writers who criticized the beast-machine hypothesis. For example, in his parodic *Voyage to the World of Cartesius*, translated into English from the original French in 1692, Gabriel Daniel, the Jesuit Father who was, Leonora Cohen-Rosenfield has argued, "the self-appointed adversary of René Descartes,"[27] has his narrator travel to the "World of Cartesius" where he is converted to Cartesianism. "My *Soul*," he states, "thus seated on the *Pineal Gland* of my Brain, as a *Queen* upon her *Throne*, to conduct and govern all the Motions of the *Machine* of my Body, was extreamly pleas'd with the change of her Ideas; and complimented her self with the honourable new Character of a *Cartesian*, wherewith I began to be distinguisht amongst the Learned." He continues:

> Before my Conversion to *Cartesianism*, I was so pitiful and Tender-hearted, that I could not so much as see a Chicken kill'd: But since I was once persuaded that Beasts were destitute both of Knowledg and Sense, scarce a Dog in all the Town, wherein I was, could escape me, for the making *Anatomical* Dissections, wherein I my self was *Operator*, without the least inkling of Compassion or Remorse; as also at the opening of the Disputes and Assemblies

[26] Nicolas Fontaine, *Mémoires pour servir à l'histoire de Port-Royal* (1738), cited in Leonora Cohen-Rosenfield, *From Beast-Machine to Man-Machine: Animal Soul in French Letters from Descartes to La Mettrie* (1940; repr., New York: Octagon Books, 1968), 54.

[27] Cohen-Rosenfield, *From Beast-Machine to Man-Machine*, 86.

of the Learn'd, which I thought good to keep at my House, for the inhanc-
ing and propagating the Doctrin of my *Master* in the Country; the first Ora-
tion I made before them, was an Invective against the Ignorance and
Injustice of that Senator, the *Areopagite,* that caus'd a *Noble Man*'s Child to be
declar'd for ever Incapacitated from entring on the Publick Government,
whom he had observ'd take pleasure in pricking out the Eyes of Jack-Daws,
that were given him to play with.[28]

Whereas Thomist arguments might make sense of the denial of entrance to
government to one who had been cruel to an animal—such a person would,
after all, be more likely to be cruel to humans—Cartesian ideas made such
a connection ridiculous. By nature humans are different from animals, and
by nature an animal cannot truly suffer pain. On this basis a human who is
cruel to an animal is, in fact, not being cruel at all. Thus fellow feeling of
humans for animals is dismissed, since humans, possessing something that
absolutely distinguishes them, can never have fellowship with animals.

The implications for understanding the nature of animals that can be
traced in Descartes's ideas, then, not only move us away from the conception of animals found in the discourse of reason; they actually make firm
what had become a fragile boundary. In the discourse of reason, with its emphasis on the empirical observation of reasonableness, a human was not necessarily born human, and once human status had been achieved, it could
easily be lost. Without reason, that discourse logically argued, a human was
not human at all but was instead a mere beast. Challenges to this discourse
emerged in the form of a return to Plutarch, and in the rediscovery of the
ideas of ancient Pyrrhonic skepticism. Montaigne's vision of animals was a
logical extension of both the problems of the discourse of reason and the
power of these alternative interpretive frameworks, and those who followed
him in England reiterated the sense of animals as exceeding the possibilities offered to them by that Aristotelian orthodoxy. It is within this context
that Descartes's ideas about animals can, perhaps, be more fully understood.
He believed the question of the nature of animals to be, as Katherine Morris has argued, "of immense importance";[29] that is, he felt it to be central to
his new science of humanity. But he was also aware of the dangers of failing
in this aspect of his thought. Without a secure statement of difference, reason could not be established as *the* means by which humans were distinguished from animals. In fact, the separation of humans and animals would
seem to be absolutely vital to the establishment of his new method.

---

[28] [Gabriel Daniel], *A Voyage to the World of Cartesius* (London: Thomas Bennet, 1692),
241–42.
[29] Morris, "*Bêtes-Machines,*" 406.

The Necessity of Cartesianism

In *Discourse on the Method* Descartes proposed that his new method would help humans to "make ourselves, as it were, the lords and masters of nature."[30] This desire for dominion was not as simple as this statement might imply, however, and in his letter to More we can see Descartes offering a vision of how it might be attained. But even as he provides absolution for the achievement of human dominion, Descartes reveals the extent to which animals had the potential to undermine that realization of power. His work shows that to believe animals could experience pain—to believe that there might be moral consequences for some of the most mundane of human actions—would be to limit human power. In this way, the beast-machine hypothesis not only offered a solution to the problematic distinction of humans from animals that had been presented by the discourse of reason, but it also and by consequence established more firmly the ways in which human superiority could be enacted in the world.

This utility of Cartesianism was reiterated in the second of Henry More's *Divine Dialogues*. Here, the Cartesian Cuphophron speaks of the "unparallel'd usefulness (in the greatest exigencies) of the peculiar notions of that stupendious wit Des-Cartes: amongst which that touching brutes being mere *machina's* is very notorious." The notoriety of the beast-machine hypothesis is acknowledged by another speaker and Cuphophron then (almost) repeats the words used by his master (Descartes) in the letter to his creator (More). Not only is the hypothesis itself notorious, he argues, but so is the usefulness of that hypothesis, "For it takes away all that conceived hardship and misery that brute creatures undergo, either by our rigid dominion over them, or by their fierce cruelty one upon another. This new hypothesis sweeps away all these difficulties in one stroke." Hylobares, "a well moralized Materialist," responds: "This is a subtil invention indeed, Cuphophron, to exclude brute creatures always from life, that they may never cease to live." Cuphophron retorts, "You mistake me, Hylobares; I exclude them from life, that they may never die with pain."[31] It is as if for this fictional Cartesian merely changing the philosophical construction of animals—altering the ways in which humans think about them—could act as an anaesthetic; as if human ideas alone created the suffering of animals. This is a kind of inverse skepticism. Whereas a skeptic like Montaigne had argued that humans create animals as lesser beings, Cuphrophon argues that humans create animals as sentient—suffering—beings.

But the reason why it was important that animals should "never die with

[30] Descartes, *Discourse on the Method*, 142–43.
[31] Henry More, "The Second Dialogue, Concerning the Providence of God," in *Divine Dialogues, Containing Disquisitions Concerning the Attributes and Providence of God* (Glasgow: Robert Foulis, 1743), 187–88.

pain" goes beyond the desire to find a way of expanding human dominion. Theodicy, the justification of the ways of God to man, also needed to be given an adequate basis from which to understand animals. As T. B., in an exchange of views published in *The Athenian Oracle* in 1703 wrote, to argue—against Descartes—that an animal had a soul would be to charge "the Almighty with injustice. Brutes have never made an ill use of their liberty, and those Natural Powers which they receiv'd in their first Creation; therefore if God punishes them with pain, and makes them not only unhappy, but equally unhappy, who are all equally innocent . . . then I can't see how God can be just." To exclude animals from life, in this analysis, is to exclude them from the problematic nature of the afterlife. T. B. goes on: "If we let go this Argument of the Mechanism of Beasts, and their final dissolution in this life, what assignable difference can there be betwixt them and Rational men?"[32] If animals suffer—feel pain—like humans, what is to say those animals do not also reap the benefits that humans can reap in the hereafter? Descartes offered a solution to these quandaries, but, as the skeptic Pierre Bayle noted, "'Tis pity that the Opinion of *Descartes* should be so hard to Maintain, and so far from Likelyhood; for it is otherwise very Advantageous to Religion, and this is the only reason which hinders some People from quitting it."[33]

The value of the beast-machine hypothesis was presented rather differently by Antoine Le Grand in 1675. He represented it in a way that made it sound like the theory actually *defended* animals: "How will *Brutes* be *Brutes* if they enjoy the use of Reason," he asked, "and have the same Sense as we; if in the same manner they *Perceive, Imagin, Judge* and *Discourse?*"[34] To argue that animals could "*Perceive, Imagin*," and so on, it seems, would itself be to deny life to animals: after all they cannot be *animals* unless they are automata because the alternative model sees a slide of animal into man which would be a destruction of the category "animal."

The utility of the beast-machine hypothesis had also been a focus in 1668, when Louis Gérard de Cordemoy defended it in yet another way, arguing that "As for me, I doubt not at all, but what hath been said of Vegetative and Sensitive Souls, which are attributed to Plants and to Beasts, hath made impious men believe, that those which are given to men, may be of the same nature."[35] On this basis—which represents the confusion as a conflation of

---

[32] T. B., *That Brutes have no Souls, but are pure Machines, or a sort of Clockwork, devoid of any sense of Pain, Pleasure, Desire, Hope, Fear, &c*, in *The Athenian Oracle*, (London: John Dunton, 1703–6), 1:504–5.

[33] Pierre Bayle, *An Historical and Critical Dictionary* (London: C. Harper et al., 1710), 4:2604–5.

[34] Anthony Le Grand, *A Dissertation Of the want of Sense and Knowledge in Brutes* (1675), in *An Entire Body of Philosophy, According to the Principles of the Famous Renate Des Cartes* (London: Samuel Roycroft, 1694), 232.

[35] [Louis Géraud de Cordemoy], *A Discourse Written to a Learned Frier, By M. Des Fourneillis; Shewing That the SYSTEME of M. DES CARTES, and particularly his Opinion concerning BRUTES,*

Platonism and Aristotelianism—Cartesianism not only solves the problem of the animal, of theodicy, of the status of the human, but it also saves philosophy from the impiety that emerges out of such a conflation. Rather than *presenting* an ethical problem, Descartes has, it seems, *solved* some of the most troubling ones.

But the emphasis that defenders placed on the devoutness of Cartesianism, on its just and truly Christian representation of the world of nature, shielded the persistent sense in which the beast-machine hypothesis was struggling to establish itself. While mechanical philosophy became the key framework for anatomical investigation, this did not automatically mean that the animal as automaton was also established as philosophical fact. In fact, in the work of a number of writers something very different can be observed.

Animal Knowledge

Marin Cureau de la Chambre, physician to the French king, for example, published in France in 1647 a text (translated into English ten years later) which offered a defense of the position of "*Porphyrius, Plutarch, Raymondus Sebondus,* for whom also *Montaigne* in his Essays hath written an Apology." These writers, Cureau's English publisher Thomas Newcomb wrote, "were al of the same opinion with our Author."[36] This is not quite true: Cureau's *Discourse of the Knowledg of Beasts* did not simply reiterate the Plutarchian position; rather, it presented to the reader a mixture of ideas from skepticism, Plutarchianism, and Aristotelianism. Cureau wrote in skeptical fashion that it was man who "hath given himself the liberty to assign to every thing the rank and order which they ought to hold in the world, and to prescribe them the function they are to exercise." Following Plutarch, he wrote that "It's impossible but we must believe, or at least suspect, that Actions which appear so reasonable, cannot but be managed by Reason." However, even in this context, it was to Aristotelianism that Cureau returned to make his case for animal capacity. His argument was

> That Beasts reason, and their reasoning is formed only of particular notions and propositions, wherein it is different from that of Men who have the faculty of reasoning universally; and that this faculty is the true difference of Man, which marks the spirituality and immortality of his soul.

---

does contain nothing dangerous; and that all he hath written of both, seems to have been taken out of the First Chapter of GENESIS (London: Moses Pitt, 1670), 48.

[36] T. N. [Thomas Newcomb], "To The Reader," in *A Discourse of the Knowledg of Beasts, wherein All that hath been said for, and against their RATIOCINATION, is Examined,* by Marin Cureau de la Chambre (London: Tho. Newcomb, 1657), n.p.

What is important here is not only the apparent paradox of Aristotelianism sitting side-by-side with skepticism and Plutarchianism. There is also the fact that the Aristotelianism that Cureau outlines fulfills the Cartesian nightmare: possession of the sensitive soul, as Descartes had argued, was offered as proof that animals could reason (albeit somewhat less reasonably than humans), even as the distinction of human and animal that was made on the basis of the relation to time was restated. Cureau was also reiterating the emphasis that the discourse of reason placed on empirically observed behavior. "Certainly," he wrote,

> there is no reasonable person who will not consent to all these truths, after having considered what Beasts do; when they see or when they hear any thing which they cannot well discern, they stay, they open their eyes and their ears, and are attentive to discover what in effect it is; for all these actions are assured marks that they doubt, and that they would assure themselves of what they do not clearly know. . . . Certainly when we see a Hare stop short at the least noise it hears, that it lifts up the head, pricks up the ears, and looks every way about it, we may assure our selves that its in trouble to know what made the noise, and that until it perceives the Huntsman, its still in doubt of what it was, and in an irresolution of what it ought to do.[37]

Cureau's return to "doubt" is surely a significant one. Whereas Descartes had proclaimed the fact that humans can doubt as one thing that could be known absolutely, Cureau trusts his eyes; he interprets the animal's pause as a moment of contemplation, and its subsequent action as evidence that a decision-making process has taken place. In Cureau's model, in fact, the animal, like Descartes, begins with doubt and ends in certainty.

But Cureau was not the only thinker to challenge the beast-machine hypothesis. Writing from exile in Paris, William Cavendish, erstwhile correspondent of Descartes's, also outlined a completely different theory of the capacity of animals. Cavendish's *General System of Horsemanship,* first published in French in 1657/58,[38] asked the obvious question of Cartesian philosophy: How can one teach a horse if it is a machine? If an animal has no capacity to think, to develop understanding over time, and if it has only a limited, not full, experience of pain, how can teaching—often involving the whip as much as the voice—work on such a creature? For Cavendish, the an-

---

[37] Cureau de la Chambre, *Discourse of the Knowledg of Beasts,* 2, 2–3, 7, and 188–89.

[38] An English "redaction" of this text, titled *A New Method and Extraordinary Invention to Dress Horses,* was published in 1667. This text, as Karen L. Raber has shown, is "entirely different" from the French original, *La méthode nouvelle.* A translation of this latter text was published by J. Brindley in 1743. See Raber, "'Reasonable Creatures': William Cavendish and the Art of Dressage," in *Renaissance Culture and the Everyday,* ed. Patricia Fumerton and Simon Hunt (Philadelphia: University of Pennsylvania Press, 1999), 43, 63n4.

swer to such a question was to assert with some certainty that Descartes was wrong.

> A horse must be wrought upon more by proper and frequent lessons, than by the heels, that he may know, and even think upon what he ought to do. If he does not think (as the famous philosopher DES CARTES affirms of all beasts) it would be impossible to teach him what he should do. But by the hope of reward, and fear of punishment; when he has been rewarded or punished, he thinks of it, and retains it in his memory (for memory is thought) and forms a judgment by what is past of what is to come (which again is thought;) insomuch that he obeys his rider not only for fear of correction, but also in hopes of being cherish'd. But these are things so well known to a complete horseman, that it is needless to say more on this subject.

The final sentence here speaks the possibility that, on a day-to-day level, humans knew that animals thought. Descartes had seen this as humans' almost-instinctive anthropomorphism, while Cavendish regards it in a more positive light. If Cavendish is right, what this shows is that the debates about reason outlined by philosophers and divines were at some distance from the lived realities of early modern men and women. Indeed, Cavendish writes that "The learned will hardly be brought to allow any degree of understanding to horses; they only allow them a certain *instinct*, which no one can understand; so jealous are the schoolmen of their rational empire."[39] Whereas Descartes argued that giving animals no mind allows for and absolves human dominion, Cavendish—skeptically—sees human dominion as merely a product of a human argument.

But Cavendish's criticism of Descartes goes further than this. As Karen L. Raber has argued, "Cavendish's more 'humane' approach [to training] in fact *humanizes* the horse much further than mere anthropomorphization might."[40] That is, in his training methods Cavendish proposes not that the horse can be understood to be like a human but that the horse *as horse* is reasonable. As Cavendish writes: "a horse's reason is to be wrought upon." This leads him to a logical conclusion: in riding "there should always be a man and a beast, and not two beasts. Indeed, a good horseman ought never to put himself in a passion with his horse, but chastise him like a kind of divinity superior to him."[41] This echoes Nicholas Morgan's emphasis on the harmonious coexistence of horse and rider, but goes further. Morgan saw the horse as natural and the rider as artful, and repeated the

---

[39] William Cavendish, *A General System of Horsemanship in all it's Branches* (London: J. Brindley, 1743), 12, 13.

[40] Raber, "'Reasonable Creatures,'" 56.

[41] Cavendish, *General System*, 13.

often-cited belief that human art is one evidence of human separation from the beast (remember Robert Gray complaining that the New World natives had "no Art, nor sciēce, nor trade, to imploy themselues, or giue themselues vnto"[42]). Cavendish, by contrast, is unwilling to make such a distinction. His horse is natural, but the discourse of reason is invoked to emphasize the possibility of human beastliness and, invoking Plutarch, to claim that the naturalness of the horse can also mean that the animal is superior to the human.

Cavendish's insistence on what Raber has termed the horse's "individual consciousness" leads to a method of training that is inevitably very different from that offered by Grisone and Blundeville, and that is even more insistent on the horse's capacity than Morgan's.[43] Whereas earlier texts argued that training a horse should begin at three and one-half years of age, Cavendish is willing to recognize the individuality of the horse: it "depends," he writes, "upon his age." But he also mentions the significance of the horse's "strength, spirit and disposition; his agility, memory, sagacity; good or bad temper; for there are horses naturally as stupid or obstinate as men." While Descartes had assumed a similarity of oyster to monkey and dog to sponge, Cavendish recognizes that there are vast differences of capacity and of character within one species.

But it is not that previous trainers have merely misunderstood the nature of horses; they have also, Cavendish argues, misunderstood how to get the best from these individuals. He argues that the trainer can use a whip as an aid during training (thus revealing the horse's capacity to feel and to learn from that—bad—feeling). But, he goes on, the trainer should only do so "sparingly. For the too frequent use of it is the cause why a horse will not go without it."[44] Rote learning is also to be avoided. Just as Owen Feltham claims that the human who learns by heart and not understanding is merely reproducing and not comprehending ideas,[45] so Cavendish believes that the horse who obeys merely by custom does not truly discern what it is obeying.[46]

Thus the horse's understanding is crucial to training. In the English *New Method and Extraordinary Invention, To Dress Horses*, the 1667 English redaction of the French *General System*, Cavendish, almost inevitably, it now seems, returns to the one obvious English example of a well-trained horse, but sees in that example evidence only of limitation: "Seeing is all the Art when they

---

[42] Robert Gray, *A Good Speed to Virginia* (London: Felix Kyngston, 1609), sig. C2ᵛ.

[43] Raber, "'Reasonable Creatures,'" 43.

[44] Cavendish, *General System*, 105, 15, 17, 19.

[45] Owen Feltham, *Resolves or, Excogitations. A Second Centvrie* (London: Henry Seile, 1628), 131.

[46] Cavendish, *General System*, 62.

teach Horses Tricks, and Gambols, like *Bankes's* Horse; and though the Ignorant admire them, yet those Persons shall never teach a Horse to go well
in the Mannage."[47] To limit your horse to learning by its senses rather than
understanding—Dekker had written, "(as *Bankes* his horse did his *tricks*)
onely by the eye, and the eare"[48]—is, for Cavendish, to limit the wonders
that that horse can perform. By implication, Morocco's fame was under-
standable—he was a wonder—but he was an unthinking, wonder. Cavendish's horses, performing their ballets, were true wonders in that they were
allowed to be thinkers.

But Descartes had attempted to counter Cavendish in a letter to him written eleven years before the publication of *The General System*. This letter, already quoted above, reveals the distance between the two men. Descartes
wrote,

> I know that animals do many things better than we do, but this does not sur-
> prise me. It can even be used to prove that they act naturally and mechani-
> cally, like a clock which tells the time better than our judgement does.
> Doubtless when the swallows come in spring, they operate like clocks.[49]

"Doubtless": this would be Descartes's ideal, his foundation for all his writing about animals, but Cavendish did doubt, did question his ideas about
the nature of animals. The beast-machine hypothesis attempted to offer humans a solution to a dangerously irresolvable problem, but instead, for
Cavendish, it merely revealed the depths of the irresolution. And Cavendish
was not alone; this is something that can also be seen in the very place where
the beast-machine hypothesis would seem to really help: science.

In the late 1640s, around the same time that Descartes and More were
exchanging views, the young Robert Boyle wrote that "certain it is, that
Beasts haue as well as we a Sence of feeling," and that "evident it is, that a
Feeling they haue of Paine." On this basis, and on the basis that an animal
does not have an immortal soul, he asked a logical and terrifying question:

> how much more misery do we inflict upon a Beaste by tormenting his Body,
> in whose delights only his Paradice consists, then in doing the Like to a Man
> <(which yet is an acknowledg'd Cruelty/> who is able to sweeten his present
> [wretchednesse] /calamity/ <sufferings> by the hope of a future, or Enjoy-
> ment of an other happynesse.[50]

---

[47] William Cavendish, *A New Method and Extraordinary Invention, To Dress Horses, And work
them according to NATURE* (1667; repr., Dublin: James Kelburn, 1740), 137.

[48] Thomas Dekker, *The Seuen deadly Sinnes of London* (London: Nathaniel Butter, 1606), 35.

[49] Descartes to Cavendish, in *Philosophical Writings*, 3:304.

[50] Robert Boyle, The Boyle Papers, Royal Society, London, vol. 37, fols. 186–93, repro-

If animals do not go to heaven—do not have the hope of a joyous afterlife—does not that mean that cruelty to animals is actually more cruel than cruelty to humans? Such a question would, inevitably, seem to challenge human dominion.

But Boyle's youthful concerns about the nature of animals and the moral status of experimentation on them were quashed because Boyle had a "change of heart," because, J. J. MacIntosh states, he "was pleased that we had animals to experiment on."[51] And we can also see how the shift from youthful anxiety to later acceptance of dominion and all that followed from it was supported by the theological framework of Boyle's experimental work. The revelations of natural truths that emerged from experimentation would, so this argument went, enhance humanity's knowledge of the world and thus of the divine. Boyle himself seemed to echo the Jacobean writer Edward Topsell when he stated in *Some Considerations Touching the Usefulness of Experimental Natural Philosophy* (1660–63) that "The two chief advantages which a real acquaintance with nature brings to our minds, are, first, by instructing our understandings, and gratifying our curiosities, and next, by exciting and cherishing our devotion."[52] To experiment on animals is, by this logic, to approach the divine. As Peter Harrison has written, "somewhat paradoxically, it was their new and more significant role as sources of arguments for the wisdom and power of God that animals great and small came to attract the unwelcome attention of anatomists, microscopists, and vivisectors."[53]

Boyle's discomfort about the experience of those animals continued, however, even as he carried on experimenting, and in 1670, some years after writing the letter on animal souls, a glimpse of his unease can be traced in his record of experiments concerning respiration. A kitten that survived suffocation—when air was removed from a "very small receiver" into which the animal had been placed—was not experimented on a second time. Instead, Boyle records that the kitten was allowed "the benefit of his good fortune," and another kitten was used in his place. Even though, as Nemesius wrote, animals are subject to chance, the idea of "good fortune" would hardly be applicable to an automaton.[54]

---

duced in Malcolm R. Oster, "The 'Beame of Diuinity': Animal suffering in the Early Thought of Robert Boyle," *British Journal for the History of Science* 22, no. 2 (1989): 173–74.

[51] J. J. MacIntosh, "Animals, Morality and Robert Boyle," *Dialogue: Canadian Philosophical Review* 35, no. 3 (1996): 462, 455.

[52] Robert Boyle, *Some Considerations Touching the Usefulness of Experimental Natural Philosophy* (1660–63), in *The Works of the Honourable Robert Boyle in Six Volumes* (London: J & F Rivington, 1772), 2:6.

[53] Peter Harrison, "Reading Vital Signs: Animals and the Experimental Philosophy," in *Renaissance Beasts: Of Animals, Humans, and Other Wonderful Creatures*, ed. Erica Fudge (Urbana: University of Illinois Press, 2004), 203.

[54] Robert Boyle, *Pneumatic Experiments about Respiration* (1670), in *The Works*, 3:360.

Boyle was not the only English scientist to continue to regard animals as sentient—and therefore pain-experiencing—beings after Descartes's hypothesis was in circulation. Henry Oldenberg, Secretary of the Royal Society, had also questioned the conclusions of Cartesians: "I cannot see how one can deduce the wise actions and strange strategems, the reflexes and feelings of pain and pleasure from these principles as they seem to do," he wrote in 1659.[55] Five years later, however, he wrote to Boyle outlining an experiment performed by Robert Hooke, in which a live dog was opened up, and

> by means of a pair of bellows (when the thorax was laid quite open, and the whole venter insimus also) and a certain cane thrust into the windpipe of the animal, the heart continued beating for a long while, at least an hour, even after the diaphragm had been cut away in great part, and the pericardium removed from the heart.

Oldenberg's report offers no insight into the pain the dog must have experienced (anesthetics were not introduced into experiments on animals until the nineteenth century), but, writing to Boyle on the same day, Hooke—the man who performed the experiment—offered a very different picture:

> I shall hardly be induced to make any further trials of this kind, because of the torture of the creature: but certainly the enquiry would be very noble, if we could find any way to stupify the creature, as that it might not be sensible, which, I fear, there is hardly any opiate will perform.[56]

Here the animal is certainly a machine worthy of examination—the enquiry is "noble"—but it is also a creature capable of experiencing the torment of the experiment. It is in the gap between these two beliefs that Oldenberg's and Hooke's anxieties can be traced. Sir John Evelyn also recorded his unease at a similar experiment in October 1667: "To Lond. . . . where was a dissection of a dog, the poore curr, kept long alive after the *Thorax* was open, by blowing with bellows into his lungs, and that long after his heart was out, and the lungs both gashed and pierced, his eyes quick all the while. This," Evelyn concluded, "was an experiment of more cruelty than pleased me."[57] For these writers, it seems, Descartes is present only in part: the dog is a ma-

---

[55] Henry Oldenberg to Saporta, 26 April 1659, in *The Correspondence of Henry Oldenberg,* trans. A. Rupert Hall and Marie Boas Hall (Madison: University of Wisconsin Press, 1965), 1:228.

[56] Henry Oldenberg to Boyle, 10 November 1664, and Robert Hooke to Boyle, 10 November 1664, in Boyle, *The Works,* 6:174, 498.

[57] Sir John Evelyn, *The Diary of Sir John Evelyn* (Oxford: Clarendon, 1955), 10 October 1667, 3:497–8.

chine, but it can also—like a human (who is, of course, also a machine in body)—experience the pain of the experiment and thus potentially make morally problematic the existence of vivisection.

But Hooke's recognition of the value of the experiment on the dog and his simultaneous desire to never repeat it amount to another way of recognizing the utility of the beast-machine hypothesis. As Daniel showed in fiction, the adoption of Descartes's ideas could remove any moral doubt about experimentation: it could get rid of the danger of animal pain. In many ways it is, as Pierre Bayle noted, strange that so few writers took up the beast-machine hypothesis. It would have indulged them, absolved them, but instead numerous humans continued to assert that animals suffered.

The continuation of the belief in animal sentience, and by consequence, the unpopularity of the beast-machine hypothesis that can be traced in many texts of the period reveal, I think, that the metaphysics of Descartes could not conquer empirical observation; that the sight of the writhing dog meant more than a philosophical assertion of absolute difference. It is, perhaps, the desire to endorse the beast-machine hypothesis in the light of this kind of resistance that leads some of Descartes's supporters to offer far more extended analyses of the status of animals than did their master. Whereas Descartes's outline of the hypothesis takes up a fairly small number of pages in relation to his work as a whole, Antoine Le Grand, to offer one example, constructs a sizeable text—almost forty folio pages—in its support. His expansion of Descartes's theory did not, however, simply offer clarification; he rendered Descartes's ideas, at times, absurd. And in doing this, he offers another view of the status of animals in pre-Cartesian thought, reiterating—and perhaps strengthening—some of the arguments I have outlined in the preceding chapters.

The Fabric of Life

For Le Grand, Descartes is "The most Illustrious CARTESIUS, who hath refined *Philosophy,* and purged it from all its folly and obscurity." He "teaches nothing but what is highly consonant to Reason [and] supposes the *Souls* of *Brutes* to be nothing else but the *Blood,* and that there is no occasion of a *Sensitive Soul* for the performance of their *operations.*" So great are the differences between the operations of humans and animals, Le Grand continues, "that it can hardly be doubted, even by the most stupid of all *Mankind,* but that they arise from different *Principles.*" One of the models that he uses to expand on the limitations of animals is familiar: he presents a living dog as a watch. But he also emphasizes the significant differences between humans' and animals' sensory capacities, and writes,

> *Brutes,* since they are void of *Mind,* and have nothing but *Corporeal Motion,* do not, for example, *see* like us, that is, as being sensible, or taking notice that they *see,* but only as we do when our *Mind* is elsewhere imployed, or called aside, tho' the *Images* of *external Objects* are painted in our *Retin Tunicle,* and perhaps also their impressions, being made in the *Optick Nerves,* dispose our *Members* to different *motions;* yet nevertheless we perceive nothing of them: In which case it appears also, that we are moved no otherwise than as *Automata,* as hath been already declared by several Examples.

In other words, while the organs of sense may be similar in humans and animals, those organs do not operate in the same way all of the time, because, when a human is applying him- or herself, those organs are joined with reason. An animal's organs never have this possibility and are therefore very different: an animal cannot reflect on what its eyes perceive; it can only perceive because, by its nature, it has no further—or higher—faculty of perception.

To reiterate this crucial difference, Le Grand presents an image remarkable in the way it absolutely removes animals from humans. He writes that the animal's memory operates "in the same manner as *folds* in *Paper* or *Linnen* render the said *Paper* or *Linnen* the more apt to readmit the same *folds* as before." To say that an animal's memory is like the "memory" of fabric is to reject out of hand the kinds of tales about animals that were used by numerous earlier scholars to argue for animal rationality. According to Le Grand's theory, the fox putting its ear to the ice in order to discern the thickness of that ice could not, in fact, discern anything and could never plan. It would be a creature merely returning to the most customary action. A curtain, after all, lacks, like the beast-machine, any kind of soul, and so does not remember to pleat itself; it merely returns to the original pleat by habit or, more accurately, by use: as Le Grand states, animals "rather *suffer* than *act.*"[58]

But Le Grand's idea of the linen-beast, while absolutely in keeping with the Cartesian denial to animals of a full experience of even the sensory world, renders almost inexplicable what it is that people see when they look at animals. Of course, the beast-machine hypothesis would answer this by noting that humans were misunderstanding the empirical, that they were— childishly—predisposed to look at animals anthropomorphically. But such Cartesian logic, although coherent, cannot fully address Cureau's deliberating hare; Cavendish's horse ballets; the pain the dog seemed to experience on Hooke's laboratory table; the anguish pictured on the face of the kitten in the image that accompanied Boyle's discussion of his experiments with the "exhausted receiver" (see figure 6.1).

[58] Le Grand, *Dissertation,* 228, 236, 230, 247, 248, 250.

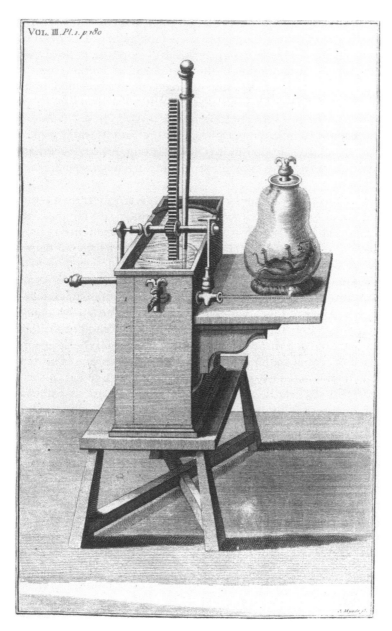

Figure 6.1. Cat in a glass receiver, from Robert Boyle, *A Continuation of New Experiments Physico-Mechanical, Touching the SPRING and WEIGHT of the AIR, and their Effects* (1669), in *The Works of the Honourable Robert Boyle* (London: J & F Rivington, 1772), volume 3, plate 1. By permission of the British Library, classmark 535i16.

This illustration, a graphic representation of an experiment described in prose, need not, of course, have represented the animal as it does. For some reason the etcher has *chosen* to depict this cat's response to the experiment as one of pain and fear, not simply one of suffocation. Perhaps this visual response gives us the truest illustration of what actually happened. In his 1670 text, *Pneumatical Experiments About Respiration*, Boyle wrote *Of the Phoenomena afforded by a new kittened kitling in the exhausted receiver.* These "*Phoenomena*" included the newly born kitten "gasp[ing] for life, and [having] some violent convulsions . . . his head downwards, and his tongue out."[59] Perhaps such an empirical record of this experiment would fail if it ignored the creature's experience of it.

There seems to be a foundation for raising a doubt, then—for being skeptical about Le Grand's analogy between linen and animals. Animals' lack of language is not merely used by Cartesians as evidence of animals' a priori lack of reason (their inability to speak reflecting their fundamental irrationality). For many Cartesians, animals' lack of language is used to *prove* lack of reason—empirical fact precedes metaphysical assertion—and lack of reason thus becomes an a posteriori fact. Thus Cartesian animals are no more *proved* to be irrational than those animals offered up by the discourse of reason—the horse that could only find its way home once on the welltrodden path, the dog that did not know it was a dog. If the assertion of animals' lack of reason requires that the animal fails to represent reason in language, then that failure, surely, can be subject to doubt.

But Le Grand's defense of the beast-machine hypothesis, while it might end with his assertion of the linen-beast, begins somewhere else. It begins, in fact, with a view of the past that reveals not only the problem that Descartes was facing but also a potential problem for historians.

Another History

Although, as I have argued in earlier chapters, the orthodox philosophical position in the early modern period was that animals lacked reason, some defenders of Descartes saw something rather different. After Descartes, it seems, the Aristotelian discourse of reason was displaced by something more theriophilic. This can be remarked at the start of Le Grand's *Dissertation Of the want of Sense and Knowledge in Brutes,* where he states,

> So far hath the Opinion concerning the *Knowledge* of BRUTE ANIMALS prevailed amongst *Men,* and so infixt hath it been in their *Minds,* that they who

[59] Boyle, *Pneumatical Experiments about Respiration* (1670), in *The Works,* 3:360. The figure of the kitten used here is from the 1772 *Works* of Boyle, but was first included in the original 1670 text.

dare think otherwise, and refuse to patronise a *Cause* which to them appears so clear, can hardly escape the censure of *Folly* and *Temerity*. In this *Opinion* almost all *Philosophers* agree, and whether induc'd by the industry and vivacity of *Sense*, which they observe some *Beasts* to be indued with, or fancying that they see some *Idea's* of *Reason* in them, they make no scruple to attribute *Knowledge* and *Ratiocination* to them, and pronounce them capable of those *perceptions* and *apprehensions*, which in reality distinguish *human kind* from all other *Creatures*, not being able to imagin how without the help of *Reason*, BRUTE ANIMALS should bring such wonderful *things* to pass, and discover in their *actings* such a world of *Ingenuity*.[60]

Belief in the "knowledge" of animals is "infixt" in human minds, and this is something that is not marginal in early modern culture but central: "almost all *Philosophers* agree."

This reading of the past is also outlined by the anonymous English reviewer of Jean Darmanson's *La Beste transformée en machine* (1684):

If any thing can mortifie the Mind of Man, it is certainly the Controversie which hath been raised not long since betwixt the *Cartesians* and other Philosophers, touching the Soul of Beasts. All Men believed, without contestation, until the time of Mr. *Descartes*, That Beasts had Knowledg. Philosophers in that had not different Thoughts from the People; they believed, as well as the Vulgar, that there was the utmost evidence for it. They only disputed among themselves, whether the Knowledg of Animals extended it self to Reason or no, and to universal Ideas? Or if it was limited by the perception of sensible Objects? Most of the ancient Philosophers have believed that Beasts reasoned; but among Christian Philosophers, the most common Opinion was the contrary. . . . [But] all Men were united in this fix'd point, and in this Article of Belief, *That Beasts have a Sentiment*. The most Subtle would have engaged, That there would never have been any Man so foolish, as to dare to maintain the contrary.[61]

The reviewer's familiar notions of "knowledg" and animal "sentiment" refer to an animal's experience of the world, and so this is something that the discourse of reason would support. However, the statement that "Most of the ancient Philosophers have believed that Beasts reasoned" sits at odds with writings by Aristotle and his followers. For them, animals were always un-

---

[60] Le Grand, *Dissertation*, 225.

[61] Anonymous review of *That Beasts are mere Machines, divided into two Dissertations: At Amsterdan by J. Darmanson, in his Philosophical Conferences in Twelves, with out the name of a Printer, 1684*, in *The Young Students Library containing extracts and abridgements of the most valuable books printed in England, and in the forreign journals, from the year sixty five, to this time* (London: John Dunton, 1692), 179.

reasonable (even though as they said this that unreason unraveled). But, perhaps the reviewer is not referring to Aristotle; perhaps this statement would be more comprehensible if we assumed that it refers to a different school of ancient thought: to Plutarch, in fact. This could be interpreted as a mistake on the reviewer's part—a denial of the true centrality of Aristotle—or as a willful *mis*representation; as an extreme outline of the past that seeks to reveal more clearly the radical nature of Descartes's work. But the emphasis on the ancient belief in animal reason could also be taken as an alternative and representative reading of a past that was marginalized in writings published during the early modern period. By placing emphasis on a thinker such as Plutarch rather than on Aristotle, this anonymous reviewer opens up the possibility that the Aristotelian discourse of reason—which represented the orthodoxy of the early modern period—did not truly reflect what "the Vulgar" believed; that, while reflecting the central concerns of key classical ideas, it failed, in fact, to make sense of the world that people lived in, and of the animals they lived with. In this way, my interpretation of pre-Cartesian ideas in England is echoed by these two texts. The equivalent binaries of human versus animal and reason versus unreason certainly existed, but they existed in constant opposition to other theories which held that animals were not only sentient but also sometimes even reasoning.

If this is true—if the representation of animals in the discourse of reason was less powerful and less widely held than has at first appeared to be the case—then it is less surprising that the beast-machine hypothesis had so few supporters. Descartes's followers were not merely challenging skepticism and Aristotelianism; they were challenging a form of what might loosely be termed Plutarchianism that was more established than is evident from contemporary writings. In fact, these Cartesian rereadings of the past may offer a glimpse of a world that is not available to us in print; may reveal the absolutely mundane nature of the belief in animal reason in the early modern period. So widely believed in, so commonly understood were the mental faculties of animals that no one bothered to write about them; as Cavendish noted: "these are things so well known to a complete horseman, that it is needless to say more on this subject." Thus, Morocco might be better interpreted not merely as a wonder but also as a simple performer. Perhaps we misinterpret the interest in the horse if we see it as focusing only on the animal's display of "intelligence." Perhaps people liked Morocco in the same way that they liked, say, Richard Tarlton: he put on a good show, made his audience laugh. And perhaps the joke is on us because we have not recognized this and are, like Aristotelian animals, focusing only on what is before our eyes—even while the horse we are looking at can see beneath the surface and tell a virgin from a whore.

Conclusion

Brutal Reasoning

To say that the beast-machine hypothesis was not taken up by many English thinkers in the early modern period and to assert the possibility that it may have been Plutarchian rather than Aristotelian ideas that reflected the way in which many humans perceived animals in that period is not, however, to deny Descartes's significance to early modern thinkers or to those who followed them. While the notion of the animal as automaton—with its assertion that animals could not fully experience pain—did not achieve dominance in the later seventeenth century, Descartes's mechanical philosophy and, most significantly here, his notion of the reasonable nature of *all* human beings did gain power. This is perhaps most clearly illustrated at the end of the eighteenth century, with the emergence of the discourse of rights in Thomas Jefferson's *Declaration of Independence* (1776) and in Tom Paine's *Rights of Man* (1792). Here the idea of the equality of all humans clearly emerges out of Descartes's ideas.[1]

But this conception of the self also brings with it the disappearance of animals from debates about the nature of the human, because animals cease to be significant once human status is established as assuredly and innately distinct from animals. The problems and anxieties that had emerged within the discourse of reason are dismissed as the human is believed complete unto itself. This assumption about the nature of the human is reflected in much modern scholarly work on the early modern period, and here we can see a contemporary place where Descartes's ideas have had an impact on human thinking about animals.

[1] The best history of this concept of the human is Tony Davies, *Humanism* (London: Routledge, 1997).

Critical Silence

In numerous modern evaluations of the early modern debates about the nature of the human, the shift from the absolute binaries human versus animal and reasonable versus unreasonable to a more ambiguous assessment of actions, and the concomitant confusion over the status of both humans and animals that exists in early modern writings, is effaced by ignoring the presence of animals or by ceasing to interpret the animals as animals. The opposition of human and animal which is the foundation of discussions of reason in the early modern period is repeated by many modern commentators with little or no analysis, as if early modern thinkers had stuck at this basic point and had not discussed it further; as if there were no real animals in early modern writings about humans. Thus in the early- to mid-twentieth century, historians of ideas writing studies of Renaissance "man"—often in order to contextualize the literature of the period—acknowledged and then veered away from the presence of animals in those early modern texts they were reading. Lawrence Babb, for example, writing in 1951, noted that in the Aristotelian model humans shared with animals the sensitive soul. Having acknowledged this, however, he made no further mention of animals in his discussion of early modern ideas about the humors and passions except to state that "Reason is the faculty which distinguishes man from beast." Following this, Babb was content to see the passionate human as engaged in "continual warfare between the rational and the sensitive, the human and the bestial, the intellectual and the physical." This warfare was an attempt by man, he wrote, to "govern his lower nature."[2] Animals have become mere metaphors of the human's own uncontrolled desires. Likewise, J. B. Bamborough in his *Little World of Man,* written a year after Babb's study, noted of Donne's poem "To Sir Edward Herbert, at Juliers" that the "bestial qualities" named in the poem must "be kept in check by the higher powers of the mind."[3] Nowhere does he comment on the possibility that the "bestial" side of humanity might actually—logically and directly—link humans with animals. Humanity's "lower nature," or what George Coffin Taylor termed in 1945 "the beast in man," otherwise known as "the evil in man's nature," is solely human; animals are absent from the discussion.[4]

    This is certainly a fair reflection of one aspect of the discourse of rea-

---

[2] Lawrence Babb, "The Physiology and Psychology of the Renaissance," in *The Elizabethan Malady: A Study of Melancholia in English Literature from 1580 to 1642* (East Lansing: Michigan State College Press, 1951), 2–3, 17, 18, 19.

[3] J. B. Bamborough, *The Little World of Man* (London: Longmans, Green, 1952), 16.

[4] George Coffin Taylor, "Shakespeare's Use of the Idea of the Beast in Man," *Studies in Philology* 42 (1945): 530.

son—it was an aspect that, as I have shown, produced a great deal of discussion and anxiety in the early modern period. But to ignore the other aspect—the link made between humans and *real* animals in many texts from the period—is to translate real animals into figurative ones in a way that is at odds with the meanings implied by early modern writers themselves. If there was a beast in man, there were also numerous beasts outside of man, beasts who not only offered ways of comprehending the (human) self metaphorically, as in Donne's poem, but also literally, as in Robert Burton's assertion of the difference between a man and a dog. Babb, Bamborough, Coffin Taylor, and numerous other mid-twentieth-century writers do not take notice of these alternatives.

In his classic study of changing human behavior norms, *The Civilizing Process* (1939), Norbert Elias seems to reflect differently than Babb, Bamborough, and others on the status of the human. Elias recognizes the emergence "since roughly the Renaissance" of "The conception of the individual as *homo clausus,* a little world in himself who ultimately exists quite independently of the great world outside."[5] This *homo clausus* was, for Elias, first established in the writings of Descartes and came to be the "Man" of late eighteenth- and early nineteenth-century liberal humanism. This Man (and initially, of course, he was male) was regarded as a being with an inalienable nature, unchanging across history. But Elias's argument is simple: *homo clausus* cannot exist if, as he argues is the case, human behavior changes over time and is therefore historical. As such Cartesian—and thus liberal humanist—Man is a myth. However, alongside this challenge to the assertion of liberal humanist ideas, Elias also reiterates a key aspect of the thinking that underpins liberal humanism. When he focuses only on the human—that is, when he ignores animals, as he does throughout his study—he reiterates the binary opposition of human and animal that was first produced in these terms by Descartes. Elias, in fact, fails to acknowledge how much the conception of the human that he takes for granted, even as he historicizes it, requires animals to exist. He does not see that his assumption of the lack of importance of animals in debates about the nature of the human is itself a form of the very humanism he is challenging.

But it is not only these old but still important texts that ignore animals. In more recent writings—writings influenced by "posthumanist" ideas[6]—a similarly narrow focus on humans can be found. Thus, even where it is critical of many of the assumptions of the liberal humanist tradition that can be

---

[5] Norbert Elias, *The Civilizing Process* (1939), trans. Edmund Jephcott (Oxford: Blackwell, 1994), 205 and 204.

[6] A useful introduction to and collection of these materials is *Posthumanism,* ed. Neil Badmington (Basingstoke: Palgrave, 2000).

traced within earlier texts, contemporary studies of the early modern period frequently remain focused, like their humanist forebears, on the human in a way that places them as post-Cartesian rather than as truly posthumanist analyses.

In one of the key modern reassessments of Renaissance selfhood, *Renaissance Self-Fashioning* (1980), for example, Stephen Greenblatt notes that Jacob Burckhardt was right to see the importance of political change for the construction of the individual in his liberal humanist *Civilization of the Renaissance in Italy* (1860). But, Greenblatt continues, Burckhardt's "related assertion that, in the process, these men emerged at last as free individuals must be sharply qualified." Greenblatt argues instead that the crumbling of "the old feudal models" created a new kind of self "precisely as a way of containing and channeling the energies which had been released." The model he presents is not of historical progress toward free individuality—is not of an emergent (pre-Cartesian) Cartesian subject—rather, it is of a new and uneasy individualism. At the end of his study Greenblatt writes of freely chosen identity: "If there remained traces of free choice, the choice was among possibilities whose range was strictly delineated by the social and ideological system in force."[7]

Greenblatt's subjects are anxious and constrained, and this is not the same as the "many-sided men" of Burckhardt's work.[8] But Greenblatt's focus is humanity; that is, he does not recognize that a key aspect of being human in the early modern period involved animals. This is because he is not applying the early modern model of the self—from Burton, Wright, and others—so much as the modern, posthuman one—from Althusser, Foucault, and others. It is because, in fact, Greenblatt is not dealing with identity so much as with subjectivity, on the (unwritten) assumption, I suggest, that animals lack subject status and therefore cannot and should not be a part of such discussions. Bruno Latour, to whose work I return later in this conclusion, sees such a division as a separation between thing (object) and citizen (subject) that sits at the heart of what he terms the "modern Constitution." This modern Constitution is one, he argues, that relies on the belief in the separation of nature from the social (human); in the distinction of the human from all other nonhumans—objects, machines, animals.[9] Thus, to return to Stephen Greenblatt, while he has shifted away from the Cartesian human that critics like Burckhardt anachronistically

---

[7] Stephen Greenblatt, *Renaissance Self-Fashioning from More to Shakespeare* (Chicago: University of Chicago Press, 1980), 161–62, 256.

[8] Jacob Burckhardt, *The Civilization of the Renaissance in Italy* (1860), trans. S. G. C. Middlemore (London: George Allen & Unwin, n.d.), 73.

[9] Bruno Latour, *We Have Never Been Modern* (1991), trans. Catherine Porter (Cambridge, Mass.: Harvard University Press, 1993), 107.

found in writings produced *before* the beast-machine hypothesis had been published, he has not gone far enough. Greenblatt's work, in fact, is part of the modern Constitution that Latour outlines; it remains, if you like, in the shadow of Descartes.

Other contemporary analyses of early modern ideas likewise place the emphasis on the human in isolation from wider networks that include animals. In *The Tremulous Private Body* (1995), his study of the emergence of the modern (Cartesian) subject, Francis Barker draws our attention to the role of Descartes in the construction of the self in what he terms "bourgeois criticism." But as well as noting the fact that this subject has a history—that it is emerging in the late-sixteenth and early-seventeenth centuries—Barker, like Greenblatt, also reiterates a key aspect of Cartesianism when he proposes that "the body has certainly been *among those objects* which have been effectively hidden from history" (emphasis mine). In his study animals are not mentioned as significant, or even as significantly absent: they are simply not there. They might be present "among those objects" hidden by the emergence of Cartesianism, but Barker does not explicitly explore this possibility, and in fact it is not clear that he is even referring to animals in this phrase at all.[10] But Barker is absolutely right on two counts in *The Tremulous Private Body:* criticism has read its post-Cartesian self back into the early modern period; and animals (if we can include them as being "among those objects") have certainly become invisible. But I would also stress that the animals that have disappeared were never simply *objects,* and that it was this fact that, in part, led to their disappearance. It was because they were so troubling, *so very like subjects,* that a new human emerged and animals vanished.

So when commentators ignore the constant presence of animals in early modern writings about the human, those commentators are—simply put—being anachronistic, as the human is being represented as explicable in and of itself. What happens, in fact, is that for modern commentators the early modern human's requirement of its animal other to establish human status is dismissed as swiftly as possible in order to present the human as whole unto itself. But what needs to be remembered is that in order to get to the point where the human species appears to be isolated, transcendent, and complete unto itself, the animal must be effaced. And this in itself means that notions of isolation, transcendence, and completeness are misinterpretations and not true reflections of what it was that early modern writers proposed. In fact, many modern critics read the Cartesian human back onto pre-Cartesian writings, even while those critics are assessing the workings of Aristotelian psychology. In doing so they implicitly and anachronistically as-

---

[10] Francis Barker, *The Tremulous Private Body: Essays on Subjection,* 2nd ed. (Ann Arbor: University of Michigan Press, 2002), 9–10.

sert that the Cartesian human is the only model of the human available. The human as a being distinct and absolutely separate from the animal is represented as a given. Many scholars would argue that this is because their focus is on humans and not animals, but I want to stress that an interest in humans in early modern writing *must* involve an interest in animals; otherwise the reading is incomplete in that it reflects ideas that emerged after Descartes and not before.

The Confused Precedent

This sense of the human that emerges in modern scholarship is not only modern, however. The absence of animals and the emphasis on the transcendent (nonbodily) human can also be traced in Sir Kenelm Digby's *Two Treatises* (1644), the first Cartesian work published in Europe. Digby, a Catholic English Royalist who spent most of the years of the Civil War and Commonwealth in exile in Paris, attempted to explain the nature of the human using Descartes's dualist ideas. Digby's dualism is reflected in the splitting of his work into two separate treatises—*A Treatise of Bodies* and *A Treatise of Mans Soule*—in which the first, while over three times the length of the second, merely establishes what is material and mortal in order to more fully comprehend the nature of the true focus of Digby's work, the immaterial and immortal nature of the human soul. What is of significance here, though, is the fact that while he proposes a Cartesian human subject—in possession of a single rational soul housed in a machine-body—Digby's animals refuse to be fully and comfortably Cartesian automata. It is the combination of these two features that, I think, makes Digby's *Two Treatises* so modern.

For Digby, "beastes . . . have a more determinate course of working, then man hath." But despite this determinate course—the limitation of the machine—"sometimes," he writes, "wee see variety." This variety of actions, which could be mistaken for free will, is, however, to be explained by bodily response: "the fume of pleasure, and the heavinesse of griefe, doe plainly shew, that the first motions do participate of dilation, and the latter of compression [of the heart]." Digby goes on, reiterating this point, arguing that "by senses, a living creature becommeth judge of what is good, and of what is bad for him." In fact, the senses operate "more perfectly" in beasts than humans because animals' "senses are fresh and untaynted, as nature made them."[11] This might sound reminiscent of Plutarch's assertion that animals

---

[11] Sir Kenelm Digby, *A Treatise of Bodies*, in *Two Treatises* (1644; repr., London: John Williams, 1645), 359–60.

have naturally superior sensory faculties to humans, but, of course, in Digby's analysis the perfection of animals is limited to the body.

Digby's desire to explain the actions of animals is given a particular impetus because of the particular nature of the actions he is considering. There are, he writes, four possible ways in which animals "seeme admirable": in the "very practice of reason, as doubting, resolving, inventing and the like"; in the "docility and practise beasts doe oftentimes arrive unto"; in their "continuate actions of a long tract of time . . . as that discourse and rational knowledge seem clearly to shine through them"; and finally in their "prescience of future events, providences, and the like." All of these possibly reasonable activities of animals are spectral presences in the discourse of reason and must be challenged if animals' status as automatons is to be fully established by Cartesianism. This is Digby's intention, and he begins by discussing the capacity for doubting that can supposedly be seen in animals.

Clearly, following Descartes's assertion of the significance of doubt—"I doubt, therefore I think, therefore I am," might be a straightforward restatement of his first principle—a doubting animal, as Cureau de la Chambre would later show, is a dangerous thing to Cartesianism. Because of this, Digby tackles the problem head-on. Animal doubt, he writes, "their long wavering sometimes between objects that draw them severall waies"—or, as we might put it, their weighing up of objects, their application of reason to those objects—is actually for Digby a natural (not reasonable) occurrence: it is "as we may observe in the sea when at the beginning of a tide of flood, it meeteth with a banke that checketh the coming in of the waves." The sea does not doubt, does not contemplate and then will itself to turn back; it merely has, by nature, a tidal motion, and this is the same with animals. As one passion pushes them in one direction (dilates the heart), so the other pulls against it (compresses the heart); there is no mind involved here—there is merely body.

Invention, which sits alongside doubting in Digby's categories of apparently reasonable actions by animals, is illustrated in part by the well-known and well-used story of the fox that feigns death in order to lure birds within its reach. Digby uses this tale because it has often been used before to prove the rationality of animals. To dismiss this reading, he returns to the old opposition of accident and decision, and argues that "chance onely doe governe their actions: and when their action proveth successfull, it leaveth such an impression in the memory, that whensoever the like occasion occurreth, that animal will follow the same method." He also notes that when these so-called reasonable actions of animals are "miscarried"—that is, when they give no impression of reasonableness—they are not written down, and so all that humans have on record are the successes, the apparent evidence of rationality. In this context an animal's reason is evidence of nothing more than a limitation in the human view of animals. It reveals, in fact, that humans are more interested in animals that behave like humans than in animals that behave like animals: the former are, no doubt, more comprehendible. Thus Digby returns to Descartes's belief in the almost instinctive anthropomorphic tendencies of humanity.

But, to return to the tale of the fox that feigns death, this, says Digby, first happens simply because the fox is so exhausted from chasing the birds that it lies down to sleep, and seeing the birds get closer and closer to him, he remains still "untill some one of them commeth within his reach, and then on a sudden he springeth up and catcheth her." There is no "designe or paines taking beforehand" in the fox's actions; rather, there is an accidental occurrence that recurs, in what appears to anthropomorphic human eyes to be a planned action, only when the fox accidentally finds himself in the same situation once again. (It is interesting that Digby does not wonder why the birds do not likewise recall from previous experience that the fox might not be dead, and keep away from him when the situation next arises. It would seem that even within this discourse there is a gradation in animal capacity.) Digby even refers back to "Montague" (Montaigne) and his argument "that Dogs use discourse," and attributes to him the tale of Chrysippus's hound. Such dogs are not syllogizing—if not that or that, then that—Digby argues; instead, their "eagernesse of hunting [has] made them overshoote the sent."[12] Again, neglect to record the occasions when this eagerness led to error would be the reason why Chrysippus's dog has remained such a popular image in philosophy. This dog is recorded because it is astonishing, because its actions look reasonable. When a dog looked as if it acted without reason, no one recorded its behavior because such actions were "normal" and expected.

In his discussion of the "docility" of animals—by which he means their capacity to be taught—Digby mentions "a baboone, that would play certaine lessons upon a guittare," apes that fetch wine from taverns for their masters, and hawks that catch their prey during the sport of hawking. He also mentions, almost inevitably, "Bankes his horse," who would, he writes, "restore a glove to the due owner, after his master had whispered that mans name in his eare." All of these activities, Digby states, seem on first sight "justly admirable," but that admiration should be tempered with a fuller understanding.

> a spectatour, who understandeth not the mystery, nor ever saw hawking before, may well admire to see a bird to dutifully and exactly obey a mans command: and may conceive she hath a reasonable soule, whereby to understand

[12] Ibid., 375, 376, 377, 379, 383.

him, and discourse of the meanes to bring his purpose to effect. Whereas indeed, all this is no more, then to make her doe for you and when you please, the same which she doth by nature to please her selfe.

Whereas Goodman had seen this as evidence of animal superiority, for Digby the hawk merely illustrates the limitations of the animal.

Likewise, activities like nest building, mentioned so regularly in Plutarchian writings, are dismissed. The birds, Digby writes, "know not what they doe whiles they build themselves houses." It is because birds do this every year—because their actions display an "invariability"—that he can state that "the birds are but materiall instruments to performe without their knowledge or reflexion, a superiour reasons counsels, even as in a clocke that is composed of severall pieces and wheeles." "The bird," he writes, "is the engine of the Artificer." Animal prescience too is interpreted as merely an act of the machine: "the foreknowing of beasts is nothing else, but their timely receiving impressions, from the first degrees of mutation in things without them; which degrees are almost imperceptible to us."[13] Animals' senses, not their powers of reason, are the real source of their foreknowledge, since, of course, they have no powers of reason.[14]

What is significant in this first Cartesian text is that many of Digby's arguments against the rationality of animal activities would not be out of place in the discourse of reason. When Digby explains the capacity of the trained hawk, for example, he argues that the hawk is not obeying human commands but is naturally acting to "please her selfe." The concept of a machine pleasing itself is, surely, an impossible one: pleasing oneself involves a notion of selfhood that Descartes would want to deny to animals. In fact, Digby's explanations of animal actions are frequently similar to those offered by Thomas Willis in his *Two Discourses Concerning the Souls of Brutes* of 1683, in which Willis refuses to accept the Cartesian view and returns to a more conventional Aristotelian argument. Willis writes:

In truth, many Actions which appear admirable in Brutes came to them at first by some accident, which being often repeated by Experience, pass into Habits, which seem to shew very much of Cunning and Sagacity; because the sensitive soul is easily accustomed to every Institution or Performance, and its Actions begun by Chance, and often repeated, pass into a Manner and Custom. So it happens sometimes by Chance among Hounds, that one had

---

[13] Ibid., 390, 393, 391, 398, 399, 418.

[14] It is just this realization that reemerged in the study of another "intelligent" horse—Clever Hans—in the late-nineteenth and early-twentieth century. See Erica Fudge, *Animal* (London: Reaktion Books, 2002), 112–16.

caught the prey, not exactly but by following a Shorter way; this Dog afterwards, as if he were much more Cunning than the rest, leaves the Hare making her turnings and windings, and runs directly to meet her another way.[15]

Willis emphasizes the existence and operation of the sensitive soul, but the solution he offers to the apparent reasoning of animals clearly echoes Digby's, even while Digby is a Cartesian.

Whether Digby merely failed to fully explicate the beast-machine hypothesis or deliberately failed to offer a comfortable outline of Descartes's idea and felt that what might be termed a "hard-line" Aristotelianism was actually a more accurate interpretation of animal actions is impossible to say. It is tempting to argue that Digby's emphasis on the empirical, on the actual activities of animals (however fabulous those actions might be), inevitably makes the application of Descartes's metaphysical, theoretical construction of animals difficult. By looking at real animals, rather than conceptualizing abstract ones, Digby frequently slides into a kind of explanatory framework that seems at a distance from the clockwork animals of Cartesianism. We can begin to see how later Cartesians such as Antoine Le Grand might have been right to note that "almost all *Philosophers* agree" that there is something like reason going on in the mind of an animal. Even Digby, the first Cartesian, found it impossible to leave such an assumption behind.

What is also clear is that, having dealt with animals in *A Treatise of Bodies,* when Digby arrives at his discussion of humans in the *A Treatise of Mans Soule,* animals have all but disappeared. This is unlike earlier analyses of human reason in which, as I have argued, animals were constantly and meaningfully present. Their disappearance in Digby's discussion of the human soul is based on Descartes's separation of body and mind, is based on Cartesian dualism. If the body is no longer a core aspect of the nature of the human (one is to be found, after all, only in one's mind) and if animals are all body, then animals are not needed to think about and define the human (and the human body is somehow not truly human). While the beast-machine hypothesis might not have become the orthodoxy in human conceptions of the animal, as even Digby's work shows, the Cartesian human has found another way to eradicate the danger of the beast.

This, it would seem to me, is closer to the model of animals that persists in the modern world. While some critics do still assert Descartes's beast-machine hypothesis as a vital and useful outline of the nature of animals,[16]

---

[15] Thomas Willis, *Two Discourses Concerning the Soul of Brutes, Which is that of the Vital and Sensitive of Man* (London: Thomas Dring, 1683), 36–37.

[16] Richard Ryder has noted that modern vivisectionists often return to a Cartesian model of the animal in which "animals do not feel pain, or do so less intensely than humans." *Victims of Science: The Use of Animals in Research* (London: National Anti-Vivisection Society, 1983),

the majority assume that animals' actions offer evidence of more than the mechanical workings of the body. But, alongside this representation of animals—that resembles the Aristotelian idea—exists a reasonable human who resembles the Cartesian model, splendid in its isolation. Hence Digby's *Two Treatises* may not only be the first Cartesian study to be published in the seventeenth century (its overall aim is Cartesian, even if it fails in its methods)—it may also be the first truly modern one, insofar as what emerges from it are invisible animals and transcendent and absolutely distinct humans.

A History of Absence

Just as many scholarly analyses of early modern ideas about the nature of the human assume a Cartesian human subject and, by doing so, render animals unnecessary, so historiography more generally can be read as a Cartesian discourse. By that I mean that at the heart of the writing of history lie certain assumptions about the nature of the subject of study, about who and what is worthy of analysis.[17] During the twentieth century, the expansion of historiographical research to include groups such as the working classes, women, ethnic others, and homosexuals greatly enriched our understanding of the past. What has persisted in these histories, however, is a focus on the human. Only toward the end of that century did scholars begin to turn away from Homo sapiens and to contemplate the role of animals in the construction of culture. And in the study of the early modern period, this approach has begun to show some of the ways in which animals have played a vital role in humans' conception of themselves. But some work simultaneously exhibits a resistance to this acknowledgment of animal centrality.

In *Inwardness and Theater in the English Renaissance* (1995), for example, Katherine Eisaman Maus, if briefly, relates her study of human interiority to questions about the nature of animals. She looks at the early modern fear of the potential for the exterior to present a false face, and notes that "it is hardly surprising that the 'problem of other minds' presents itself to thinkers and writers not so much as a question of whether those minds exist as a question of how to know what they are thinking." The text she turns to here is the Raleigh-attributed *Sceptick*, and she notes that the desire to explore the (possible) consciousness of animals is a logical one and a very real one in

12. For a modern defence of Descartes see Peter Harrison, "Do Animals Feel Pain?," *Philosophy* 66, no. 1 (1991): 25–46, and Harrison, "Descartes on Animals," *Philosophical Quarterly* 42 (1992): 219–27.

[17] On this issue as it speaks to questions of biography in particular see Erica Fudge, "Animal Lives," *History Today* 54, no. 10 (2004): 21–26.

this period.[18] It is perhaps telling that Maus does not expand on this point in her study: she notes the early modern interest in animals but leaves it at that. The animal is made visible only to disappear once again.

In Michael C. Schoenfeldt's *Bodies and Selves in Early Modern England* (1999), a study of the role of physiology in the construction of the self in early modern writing, the author challenges any simplistic desire to place the early modern self simply in the mind and rightly notes the difficulty of such a conception of selfhood "for those of us who are the inheritors of the Cartesian philosophical tradition." What he also notes is that the relegation of animals to the realm of the symbolic does not reflect accurately the nature of the discussions in the texts he analyses. For example, in his study of *The Faerie Queene* he notes that, for Belphoebe, "beasts, of course, are both external threats and conventional representatives of the internal passions that she so expertly controls in her deliberate and active virginity."[19] These are simultaneously real animals and the bestial desires of the individual. Schoenfeldt's discussion, however, goes no further.

It might seem unfair to criticize these texts for not taking more interest in and notice of animals when their true focus is on humans. Why should one expect a critic to write about something that is so apparently peripheral to their work? My answer to this question is one I have already given: thinking about humans in the early modern period *is* thinking about animals. There should be little separation for us between humans and animals because there was so little separation for those writers we are interpreting. I want to take this further: to think only about humans when reading early modern texts is to apply anachronistically Cartesian ideas to pre-Cartesian thought. Barker, Maus, Schoenfeldt, and others note the significant differences between early modern and modern concepts of identity, but despite noting the (possible) role of animals in earlier conceptualizations, they continue to focus only on humans.

But as our interest in the environment and in animal welfare develops in contemporary Western thought (for pressing reasons), so a historical dimension is being added to those modern debates, and so animals begin, inevitably, to emerge in historical analysis. In her discussion of the early modern humors, for example, Gail Kern Paster turns to animals and recognizes the importance of taking them seriously if humans are themselves to be taken seriously. Paster's conception of the shift in the construction of the human is not from pre-Cartesian to post-Cartesian ideas but from

[18] Katherine Eisaman Maus, *Inwardness and Theater in the English Renaissance* (Chicago: University of Chicago Press, 1995), 7.

[19] Michael C. Schoenfeldt, *Bodies and Selves in Early Modern England: Physiology and Inwardness in Spenser, Shakespeare, Herbert, and Milton* (Cambridge: Cambridge University Press, 1999), 10, 45.

what she terms pre-Enlightenment to post-Enlightenment ones. But what emerges from Paster's reading of this shift is a disappearance of the animal. Moods experienced by Shakespearean characters are not, she argues, "disembodied mental event[s]" (such would be anachronistic) but are comprehendible through an "analogical network" that includes animals. She argues that the representation of animals as sharers with humans of humors such as melancholy in early modern discussions is not to be understood as "anthropomorphism—the attribution of complex emotional states and cognitive operations to animals"; rather, it is "a significant aspect of the period's deepest habits of thought, which involve a specific placing of self within a universe understood to be filled with desire and moved by the strivings of appetite."[20] Animals are included in such discussions—and are included *as analogous with humans*—because that it how they were understood.

Paster's discussion, then, is an important reminder of the ways in which early modern ideas about the shared physiology of humans and animals fed into notions of shared psychology, but she does not look in any detail at the role of animals in discussions of reason. This is also true of my earlier work *Perceiving Animals*. I now realize that much of what I illustrated in that book concerned how the discourse of reason was articulated in various areas of early modern culture—how the discourse was used to construct the human in humanist, religious, scientific, legal, and political realms of life.[21] A corrective to this absence of animals can be found in the first contemporary study that makes the shifts in focus from human to animal truly visible: Keith Thomas's *Man and the Natural World* (1983). This is the most exhaustive survey of early modern ideas about animals in current historiography, but still, I would argue, it fails to fully explore some of the more problematic resonances of many of the writings which are part of what I have termed the discourse of reason. Thomas writes, for example, that "By embodying the antithesis of all that was valued and esteemed, the idea of the brute was as indispensable a prop to established human values as were the equally unrealistic notions held by contemporaries about witches or Papists."[22] This is certainly true on one level, but Thomas's interrogation of the materials he refers to does not go far enough. I would argue that in representing animals as "props" of human values, these animals come, paradoxically, to undo those very values. If it is logical within the discourse of reason to suggest, as Burton does, for example, that a man lacking in self-knowledge does not

---

[20] Gail Kern Paster, *Humoring the Body: Emotions and the Shakespearean Stage* (Chicago: University of Chicago Press, 2004), 142, 137, 187.

[21] Erica Fudge, *Perceiving Animals: Humans and Beasts in Early Modern English Culture* (Basingstoke: Macmillan, 2000).

[22] Keith Thomas, *Man and the Natural World: Changing Attitudes in England, 1500–1800* (London: Penguin, 1984), 40.

know the difference between himself and a dog, how can that dog possibly be upholding difference?

Bruce Boehrer corrects Thomas's emphasis on the orthodoxy of the discourse of reason by exploring texts by Montaigne. Classifying different constructions of humans and animals in early modern writings under the terms "absolute anthropocentrism," "relative anthropocentrism," and "anthropomorphism," Boehrer, if briefly, draws out many of the different and paradoxical ways in which animals and humans were defined in early modern writing.[23] He looks at texts from what I term the discourse of reason, as well as writings by Plutarch, but his more detailed discussions focus on play texts, and this, I think, means that the kind of emphasis that early moderns placed on what animals actually did—on empiricism—is missing. Apart from a few notable exceptions, animals in the drama of the period were not *seen* on stage.[24] Because of this, Boehrer does not recognize fully the context into which we should place the Cartesian "beast-machine" in seventeenth-century ideas.

But these attempts by Paster, Thomas, Boehrer, and myself to shift attention to the animal alongside (as inseparable from) the human remain unorthodox, because what is orthodox in historiography is the question of agency—by which I mean here the self-willed direction of actions toward a desired objective—is the distinction, as Bruno Latour argues, between citizen and thing. Just as, for Sir Kenelm Digby, animals' actions are merely instinctive (birds "know not what they doe whiles they build themselves houses") so they are excluded from historical analysis by many contemporary historians because such a conception of agency remains central in their discourse.

But there is an alternative way of thinking about the subject of history. If the emphasis on agency in history seems to be an emphasis on the human—for agency is, in this context, human, even when that human is perceived to be limited by circumstances—any challenge to this emphasis might seem to suggest that the alternative organizing principle of historiography is to turn instead to chance, to what Nemesius argued governs animal actions. But this is not the case. A broader notion of agency might allow us a way of rethinking not only how we conceptualize the arrangements of culture and the structures of thought that organize humans' perception of animals and of themselves in the past—it might also allow us to rethink how it is that we understand the history of being human, and from that gain a better understanding of what it means to be a human now.

[23] Bruce Boehrer, *Shakespeare Among the Animals: Nature and Society in the Drama of Early Modern England* (New York and Basingstoke: Palgrave, 2002), 1–40 and passim.

[24] On animals on the early modern stage see Louis B. Wright, "Animal Actors on the English Stage before 1642," *PMLA* 42 (1927): 656–69; and Matthew Bliss, "Property or Performers? Animals on the Elizabethan Stage," *Theatre Studies* (1994): 45–58.

Animal Agents

What does it mean to assert the possibility that human agency might not, then, be the only motivating force of the past? Or, to word it somewhat differently, what does it mean to propose, as I do, that animals might also have played a role in the construction of the world explored by historians? How can such a suggestion be made without reverting to anthropomorphism? without proposing that animals—beings without the capacity to record for us their own desires—are just like humans? We can turn to some of the key ideas raised by a theory that proposes another way of thinking about the ways in which we structure, and hence understand, the world. We can turn to actor-network theory.

One of its main thinkers, Bruno Latour, outlines key assumptions of what he terms the "modern Constitution" that echo Cartesian notions. (Descartes is only mentioned in passing in Latour's study.) Latour argues that the modern Constitution is paradoxical; it consists of two parts that coexist but simultaneously stand at odds with each other. The first is the work of "translation" by which, as Philip Armstrong has put it, "the endless and occult permeation between the natural and the social, the non-human and the human domains" takes place.[25] The second aspect of the modern Constitution stands absolutely against this recognition of the intertwining of the natural and the social and is, for Latour, the work of "purification" whereby the dichotomy of nature and the social is established and maintained. This separation is also between "human beings on the one hand [and] nonhumans on the other." Latour continues, "Without the first set, the practices of purification would be fruitless or pointless. Without the second, the work of translation would be slowed down, limited, or even ruled out." What is crucial for the modern Constitution is that we live as if nature and the social were, like nonhumans and humans, in absolute opposition, while ignoring the proliferation of hybrids. But Latour goes on: "As soon as we direct our attention simultaneously to the work of purification and the work of hybridization, we immediately stop being wholly modern, and our future begins to change."[26] It is by acknowledging that human beings are inextricably linked with nonhumans—animals, objects, machines, "things"—that a new way of looking at the world, and the possibility of a new future, emerges.

What is established by the acknowledgment of the link between the natural and the social, Latour argues, is a "Parliament of things."[27] But this

---

[25] Philip Armstrong, "What Animals Mean, in *Moby-Dick,* for example," *Textual Practice* 19, no. 1 (2005): 95.

[26] Latour, *We Have Never Been Modern,* 10–11.

[27] Ibid., 144.

leads not to a degradation of the human; rather, it is an establishment of a new being. Latour writes, "the human, as we now understand, cannot be grasped and saved unless that other part of itself, the share of things, is restored to it. So long as humanism is constructed through contrast with the object that has been abandoned to epistemology, neither the human nor the nonhuman can be understood." The notion that a shift from an assumption of the human existing in isolation to a celebration of human hybridity will destroy humanity is itself a product of the work of purification, but Latour reminds us that in this new world of networks "nothing is sufficiently inhuman to dissolve human beings in it and announce their death." The human is always already enmeshed in the nonhuman to the extent that humanity—what is valued by the modern Constitution—does not actually exist outside of that Constitution. In the new Constitution that celebrates hybridity, "History is no longer simply the history of people, it becomes the history of natural things as well."[28]

In his 1992 outline of this theory John Law argues that "Machines, architectures, clothes, texts—all contribute to the patterning of the social. And—this is my point—if these materials were to disappear then so too would what we sometimes call the social order." He goes on to explain that actor-network theory "says that there is no reason to assume, *a priori,* that *either* objects *or* people in general determine the character of social change or stability." The theory, therefore, "denies that people are *necessarily* special."[29] It is not difficult to see how this might allow for an understanding of the ways in which animals impact on change. In the discourse of reason the human is certainly only fully comprehensible when read within a network that includes animals, and so animals have the capacity to determine the nature of the human as much as the human determines that of the animal. In this we can see how animals can be said to possess agency. Moreover, as I stated earlier in the book, if animals disappear, if humans cease to have animals to think with, live with, and observe, then the human will disappear. Accordingly, the a priori status of the human as *the* reasonable being cannot hold up.

What my work here also shows, I hope, is that the discourse of reason had no real claim to truth even in a period when it remained the orthodox basis for the construction of human–animal relations. The degeneration of the discourse becomes visible when the inconsistencies inherent in that discourse are confronted by the reemergence of skeptical ideas and the increasing emphasis on empiricism in humanity's perception of animals in

---

[28] Ibid., 136–37, 82.
[29] John Law, "Notes on the Theory of the Actor-Network: Ordering, Strategy, and Heterogeneity," *Systems Practice* 5 (1992): 382, 383.

written (and therefore "serious") debate. In fact, according to Cartesian scholars themselves, the reemergence of skepticism and the preponderance of empiricism in intellectual debate may have simply taken that intellectual debate closer to the popular one in which King James VI and I and Morocco's human audience knew that their animals had some kind of intellectual capacity. In the Cambridge University debate about the sagacity of dogs, you may recall, the orthodox argument could only read King James's hounds as symbols of his regal status and not as real dogs; such was the desire to maintain the ordered separation that the discourse of reason seemed to require. At this point, the ordered separation ceased to function so simply and so monolithically, and what emerged was Montaigne's question ("When I play with my cat, who knows if I am not a pastime to her more than she is to me?"), is James VI and I's statement ("I had myself . . . a dog") and Richard Tarlton's exclamation ("*God a mercy Horse*"). What also emerged, somewhat later and even more powerfully, was Descartes's beast-machine hypothesis; a new order in which the dangers to the status of the human that came from the reemergence of skepticism and the emphasis on real animals was offset by the disappearance of those animals. No longer thinking, feeling subjects, Descartes's animals were the antithesis of the human: not merely silent, but silenced.

But, of course, Descartes's writings and the Cartesianism that emerged after him are themselves attempts to present as natural and true what is, in fact, a product of human failure and human desire. Cartesianism's greatest success, perhaps, was that it turned "human" from a verb into a noun,[30] making reason the inherent possession of the individual, rather than the product of actions or of a network that relies on and includes animals. The true success of Cartesian theory, then, is that it eliminated animals from the picture and thus made them seem inconsequential. Its success was so great that animals are still almost wholly eliminated from discussion.

However, as with the power of the discourse of reason, the taken-for-grantedness of the existence of the Cartesian human is precarious, and it is by bringing animals back into the picture that that taken-for-grantedness may be challenged. It is by recognizing the real—and active—role that animals have played in the construction of this being called the human that we can challenge Cartesianism and the worldview that follows from it, in which animals are relegated to object status (even if we do not think of them as machines). This is not only a question of historiography but also a question of ethics. In fact, in writing histories without animals, we continue to make natural the ways of thinking that efface those animals, and we hide the

---

[30] This is an adaptation of Law's description of the social structure. See Law, "Notes on the Actor-Network," 385.

fact that the production of meaning and order is the work of many, and not always human, agents. In this way, as well as ignoring animals, we not only misrepresent ourselves and our pasts but limit our possible futures as well.

This book began, as I have said, as an attempt to outline some early modern assumptions about animal rationality. What I discovered was that to do this I had to follow the trajectory that early modern writers followed and look at humans as well as—if not more than—animals. In doing this, what emerged was not simply a history of a debate about animal capacity; it was also a history in which the construction of the modern human was being played out. The debates in the early modern period—Aristotelian, Plutarchian, skeptical, Cartesian—not only offer very different assessments of the capacity of animals but also present outlines of the human that reveal neither a simple assertion of selfhood in the early modern period, nor a struggle to assert selfhood in the face of social and ideological constraints. What is revealed in all of these different ways of thinking is the centrality and sometimes the danger of animals.

Animals were needed in order to express the superiority of humans—to place men next to angels, in fact. But by holding this position, animals constantly undermined ideas of difference and superiority. In addition, a growing interest in empirical observation revealed to many early modern writers a capacity in animals that was not addressed by the orthodox discourse. Instead of retreating from this anomaly, however, many writers took the anomalous nature of animals as their subject, giving rise to studies finding not irrationality, but the possibility of a dangerous rationality, in animal actions.

This would seem, perhaps, simply a quaint history, a—hopefully—interesting piece of research if it were not for the fact that at the end of the period that I have focused on, a model of the human arose that not only attempts to banish animals from the debate but also views superiority as something innate rather than achieved. While one might see something potentially meritocratic in the emphasis on the actions of the individual that are the focus of empiricism (although it was not, of course, intended to be that), the Cartesian assertion of human difference makes that merit species-centric; makes animals, by the mere fact of their existence, inferior. This model has persisted into the twenty-first century.

By following the history of how this model of human–animal relations developed—by revealing it, in fact, to have a history—I hope that it may become possible to reassess not only the early modern period itself but also our own. And by doing this, to think again about those vital beings, animals, and the significant and productive roles they have played in our past, play in our present, and will continue to play in our futures. At the very least, I hope that this study has shown that the assumption that animals lack reason was not held without difficulty in early modern England; and that that assumption contained within itself the seeds of its own undoing. What I also hope to have shown is that it might well be philosophically rigorous to assert, with King James, "I had myself a dog." Still, the orthodox context of philosophical debate may well undercut such an intervention by labeling it merely anecdotal, even though that orthodoxy's own assertions, as skepticism has shown, can themselves be dismissed as merely theoretical. Given the orthodoxy's difficulties, the simple claim that an animal is not reasonable—while a human is—ought to be rather difficult to make. And perhaps the hierarchies that follow will likewise begin to come unstuck. Perhaps Richard Tarlton was right: "God a mercy horse indeed."

Primary Materials

Adams, Thomas. *Mystical Bedlam, or The World of Mad-Men*. London: George Purslowe, 1615.

——. *The Workes of Tho: Adams*. London: Tho. Harber, 1629.

Aelian. *On the Characteristics of Animals*. Translated by A. F. Scholfield. London: Heinemann, 1958.

Agrippa, Henry Cornelius. *Of the Vanitie and vncertaintie of Artes and Sciences: Englished by Ia. San. Gent*. London: Henrie Bynneman, 1575.

Anon. *The Contention Betweene three Brethren*. London: Robert Raworth, 1608.

Anon. *Don Zara Del Fogo: A Mock-Romance*. London: Tho. Vere, 1656.

Anon. *The Sceptick*. In William M. Hamlin, "A Lost Translation Found? An Edition of *The Sceptick* (c. 1590) Based on Extant Manuscripts [with text]." *English Literary Renaissance* 12, no. 2 (2001): 34–51.

Anon. *A Strange and Miracvlovs Accident happened in the Cittie of Purmerent, on New-yeeres euen last past 1599*. London: John Wolfe, 1599.

Anon. Review of *That Beasts are mere Machines, divided into two Dissertations: At Amsterdan by J. Darmanson, in his Philosophical Conferences in Twelves, with out the name of a Printer, 1684*, in *The Young Students Library*, 179–82. London: John Dunton, 1692.

Anon. *A Trve and Sincere declaration of the purpose and ends of the Plantation begun in Virginia*. London: I. Stepneth, 1610.

Apuleius. *The Golden Asse of Lucius Apuleius*. Translated by William Adlington. 1566. Reprint, London: Simkin Marshall, 1922.

Aquinas, Thomas. *Summa Theologiae*. Translated by R. J. Batten. London: Blackfriars, 1975.

Aristotle. *The Works of Aristotle*. 2 vols. Chicago: Encyclopædia Britannica, 1952.

——. *De Memoria et Reminiscentia*. In *Aristotle on Memory*, by Richard Sorabji. London: Duckworth, 1972.

——. *The Problemes of Aristotle, with other Philosophers, and Phisitions.* London: Arnold Hatfield, 1597.

Artimedorus. *The Iudgement, or exposition of Dreames, Written by Artimodorus, an Auntient and famous Author, first in Greeke, then Translated into Latin, After into French, and now into English.* London: William Jones, 1606.

Augustine, St. *Against the Academics.* Translated by John J. O'Meara. London: Longmans, 1951.

——. *The Greatness of the Soul.* Translated by Joseph M. Colleran. Westminster and London: Newman Press and Longmans, 1949.

B., R. *Curiosities: Or The Cabinet of Nature.* London: N. and I. Okes, 1637.

B., T. *That Brutes have no Souls, but are pure Machines, or a sort of Clockwork, devoid of any sense of Pain, Pleasure, Desire, Hope, Fear, &c.* In *The Athenian Oracle.* Vol. 1. London: Andrew Bell, 1703.

[Baldwin, William]. *A treatye of Moral Philosophy containing the sayinges of the wise.* London: Rycharde Tottill, 1564.

Ball, Thomas. *The Life of the Renowned Doctor Preston, writ by his pupil, Master Thomas Ball, D. D. Minister of Northampton, In the Year 1628,* ed. E. W. Harcourt. Oxford: Parker, 1885.

[Bastard, Thomas]. *Chrestoleros. Seuen bookes of Epigrammes written by T. B.* London: I. B., 1598.

Bateman, Stephen. *Batman vppon Bartholome, his Booke De Proprietatibus Rerum.* London: Thomas East, 1582.

Bayle, Pierre. *An Historical and Critical Dictionary.* 4 vols. London: C. Harper et al., 1710.

Blundeville, Thomas. *The Fower chiefyst offices belongyng to Horsemanshippe.* London: William Serres, 1565.

Boaystuau, Pierre. *Theatrum Mundi, The Theatre or rule of the world.* Translated by John Alday. London: Thomas Hacket, ?1566.

Boyle, Robert. *The Works of the Honourable Robert Boyle in Six Volumes.* London: J. & F. Rivington, 1772.

The Boyle Papers, Royal Society, London. Vol. 37, fols. 186–93. Reproduced in "The 'Beame of Diuinity': Animal suffering in the Early Thought of Robert Boyle," by Malcolm R. Oster. *British Journal for the History of Science* 22, no. 2 (1989): 151–80.

Brathwait, Richard. *A Strappado for the Diuell.* London: Richard Redmer, 1615.

——. *The English Gentleman.* London: Iohn Haviland, 1630.

——. *A Strange Metamorphosis of Man, transformed into a Wildernesse.* London: Thomas Harper, 1634.

Browne, Sir Thomas. *The Major Works.* Edited by C. A. Patrides. London: Penguin, 1977.

Bryskett, Lodowick. *A Discovrse of Civill Life.* London: Edward Blount, 1606.

Burton, Robert. *The Anatomy of Melancholy.* Oxford: Henry Cripps, 1624.

Caius, John. *Of Englishe Dogges, the diuersities, the names, the natures, and the properties.* Translated by Abraham Fleming. London: Richard Johnes, 1576.

Calvin, John. *Institutes of the Christian Religion.* Translated by Henry Beveridge. 2 vols. London: James Clarke, 1949.

Cardano, Girolamo. *Cardanus Comforte translated into Englishe*. London: Thomas Marshe, 1573.

Carew, Richard. *The Survey of Cornwall*. London: SS, 1602.

Caussin, Nicholas. *The Holy Covrt, or The Christian Institution of Men of Qvality*. Translated by T. H. [Thomas Hawkins]. Paris: St. Omer, 1626.

Cavendish, William. *A Genoral System of Horsemanship in all it's Branches*. London: J. Brindley, 1743.

———. *A New Method and Extraordinary Invention, To Dress Horses, And work them according to NATURE*. 1667. Reprint, Dublin: James Kelburn, 1740.

Cevolus, Francis. *An Occasionall Discourse, upon an Accident which befell his MAIESTY in hunting*. London: Iohn Norton, 1635.

Chamberlain, Robert. *Nocturnall Lucubrations: Or Meditations Divine and Morall*. London: M. F. for Daniel Frere, 1638.

Chambre, Marin Cureau de la. *A Discourse of the Knowledg of Beasts, wherein All that hath been said for, and against their RATIOCINATION, is Examined*. London: Tho. Newcomb, 1657.

Charron, Pierre. *Of Wisdome, Three Bookes Written in French by Peter Charro[n] Doct. of Lawe in Paris*. Translated by Samson Lennard. London: Edward Blount and Will. Aspley, 1607.

Cicero. *The Academics of Cicero*. Translated by James S. Reid. London: Macmillan, 1880.

Cleaver, Robert. *A Plaine and Familiar Exposition of the Eleventh and Twelfth Chapters of the Proverbes of Salomon*. London: Richard Bradocke, 1608.

Coeffeteau, F. N. *Table of Humane Passions. With their Causes and Effects. . . . Translated into English by Edw: Grimeston Sergiant at Armes*. London: Nicholas Okes, 1621.

Cogan, Thomas. *The Haven of Health*. London: W. Norton, 1584.

[Cordemoy, Louis Géraud de]. *A Discourse Written to a Learned Frier, By M. Des Fourneillis; Shewing That the SYSTEME of M. DES CARTES, and particularly his Opinion concerning BRUTES, does contain nothing dangerous; and that all he hath written of both, seems to have been taken out of the First Chapter of GENESIS*. London: Moses Pitt, 1670.

Cornwallis, Sir William. *Essayes*. London: I. Windet, 1610.

Crashaw, William. *A Sermon Preached in London before the right honourable the Lord LAWARRE, Lord Governour and Captaine Generall of Virginia*. London: William Welby, 1610.

Dando, John, and Harry Runt. *Maroccus Extaticus. Or, Bankes Bay Horse in a Trance*. London: Cuthbert Burby, 1595.

[Daniel, Gabriel]. *A Voyage to the World of Cartesius*. London: Thomas Bennet, 1692.

Davies, John. *Epigrammes*. Middleburgh, ?1590.

———. *Nosce Teipsum*. London: Richard Field, 1599.

Dekker, Thomas. *The Seuen deadly Sinnes of London*. London: Nathaniel Butter, 1606.

Descartes, René. *The Philosophical Writings of Descartes*. Translated by John Cottingham, Robert Stoothoff, and Dugald Murdoch. 3 vols. Cambridge: Cambridge University Press, 1985.

Digby, Kenelm. *Two Treatises: In the one of which, The Natvre of Bodies, In the other the Nature of Mans Soule is looked into*. London: Iohn Williams, 1645.

Diogenes Laertius. *Lives of Eminent Philosophers.* Translated by R. D. Hicks. London: William Heinemann, 1925.

Dod, John, and Robert Cleaver. *A Godly Forme of Houshold Gouernment, For the ordering of priuate Families, according to the direction of Gods Word.* London: Thomas Man, 1630.

———. *A Treatise of the Exposition Vpon the Ten Commandments.* London: Thomas Man, 1603.

Donne, John. *Sermon of Valediction at his Going into Germany Preached at Lincoln's Inn April 18, 1619.* London: Nonesuch Press, 1932.

———. *The Complete English Poems.* Edited by A. J. Smith. 1971. Reprint, London: Penguin, 1986.

Du Bartas, Guillaume de Salluste. *Bartas His Deuine Weekes & Workes.* Translated by Joshuah Sylvester. London: H. Lownes, 1605–6.

Duvair, Guillaume. *The Moral Philosophy of the Stoicks, Written in French, and englished for the benefit of them which are ignorant of that tongue. By T. I. Fellow of New Colledge in Oxford.* London, Felix Kingston, 1598.

Elviden, Edmond. *The Closet of Counsells, conteining The aduice of diuers wyse Philosophers, touchinge sundry morall matters.* London: Thomas Colwell, 1569.

Estienne, Henry. *Maison Rustique, or The Countrie Farme.* London: Edm. Bollifant, 1600.

Feltham, Owen. *Resolves or, Excogitations. A Second Century.* London: Henry Seile, 1628.

[Fiston, William]. *The Schoole of good manners.* London: I Danter for William Ihones, 1595.

Fitz-Geffry, Charles. *Compassion Towards Captives.* Oxford: Leonard Lichfield, 1637.

Franzius, Wolfgang. *The History of Brutes; Or, A Description of Living Creatures.* 1612. Rendered into English by N. W. London: E. Okes, 1670.

Gascoigne, George. *A Delicate Diet, for daintie mouthde Droonkardes.* 1576. Reprint, London, 1789.

Goodman, Godfrey. *The Fall of Man, or The Corrvption of Nature, Proved by the light of our naturall Reason.* London: Felix Kyngston, 1616.

———. *The Creatvres Praysing God: Or, The Religion of dumbe Creatures.* London: Felix Kingston, 1622.

Gouge, William. *Of Domesticall Duties, Eight Treatises.* London: Edward Brewster, 1634.

Gratorolus, Gulielmus. *The Castel of Memorie . . . Englished by Willyam Fulwood.* London, Rouland Hall, 1562.

Gray, Robert. *A Good Speed to Virginia.* London: Felix Kyngston, 1609.

Greenham, Richard. "Of Joy and Sorrow." In *The Workes of the Reverend and Faithfull Servant of Iesus Christ M. Richard Greenham.* London: Felix Kyngston, 1601.

Grotius, Hugo. *The Illustrious Hugo Grotius of the Law of Warre and Peace.* London: T. Warren, 1654.

Guillimeau, James. *Child-Birth Or, The Happy Deliverie of Women.* London: A. Hatfield, 1612.

Hall, Joseph. *The Arte of Divine Meditation: Profitable for all Christians to know and practice.* London: H. L., 1607.

——. *Meditations and Vowes Diuine and Morall; Seruing for direction in Christian and Ciuill Practice: Diuided into two Bookes.* London: Humfrey Lownes, 1607.

——. *Characters of Vertues and Vices in two Bookes.* London: Melch. Bradwood, 1608.

——. *Salomons Diuine Arts, of 1. Ethickes, 2. Politickes, 3. Oeconomicks. That is; the Gouernment Of 1. Behaviour, 2. Common-wealth, 3. Familie. Drawne into Method, out of his Prouerbs & Ecclesiastes.* London: H. L. 1609.

——. *The Works of Joseph Hall Doctor in Diuinitie, and Deane of Worcester.* London: Thomas Pauier, 1625.

——. *Occasionall Meditations.* Set forth by R[obert] Hall. London: Nath: Butter, 1630.

——. *Virgidemiarum: Satires in Six Books.* 1597. Reprint, Oxford: R. Clements, 1753.

Hamor, Ralph. *A True Discourse of the Present Estate of Virginia, and the success of the affaires there till 18 June 1614.* London: W. Welby, 1615.

Harington, John. *The Most Elegant and Witty Epigrams of Sir John Harington, Knight.* London: G. P., 1618.

Harrison, William. *An Historicall Description of the Islande of Britayne,* in Raphael Holinshed, *The Firste volume of the Chronicles of England, Scotlande, and Irelande.* London: Iohn Harrison, 1577.

Hazlitt, W. Carew, ed. *Shakespeare Jest-Books: Reprints of the Early and very Rare Jest-Books supposed to have been used by Shakespeare.* 3 vols. London: Willis & Southeran, 1864.

[Henshaw, Joseph]. *Meditations miscellaneous, Holy and Humane.* London: R. B., 1637.

Herbert, Edward, Lord of Cherbury. *De Veritate.* Translated and with an introduction by Meyrick H. Carré. Bristol: University of Bristol, 1937.

Heywood, Thomas. *Gynaikeion: or, Nine Bookes of Various History Concerninge Women; Inscribed by name of ŷ Nine Muses.* London: Adam Islip, 1624.

——. *Philocothonista, or The Drvnkard, Opened, Dissected, and Anatomized.* London: Robert Raworth, 1635.

Hill, Thomas. *The Moste pleasaunte Arte of the Interpretacion of Dreames.* London: T. Marsh, 1576.

Hill, William. *The Infancie of the Soule: Or, The Soule of an Infant.* London: C. Knight, 1605.

Homer. *The Odyssey.* Translated by George Chapman. 1614–15. Reprint, Ware: Wordsworth, 2002.

Huarte, Juan. *Examen de Ingenios. The Examination of mens Wits.* London: Adam Islip, 1596.

Iamblichus. *Iamblichus' Life of Pythagoras, or Pythagoric Life.* Translated by Thomas Taylor. 1818. Reprint, London: John M. Watkins, 1926.

Jewel, William. *The Golden Cabinet of true Treasure: Containing the summe of Morall Philosophie.* London: John Crostley, 1612.

Jones, Richard. *A Briefe and Necessarie Catechisme.* London: Thomas East, 1583.

Jonson, Ben. *Ben Jonson: The Complete Poems.* Edited by George Parfitt. London: Penguin, 1988.

Jorden, Edward. *A Briefe Discourse of a Disease Called the Suffocation of the Mother.* London: John Windet, 1603.

Joubert, Laurent. *Treatise on Laughter* (1579). Translated by Gregory David De Rocher. Tuscaloosa: University of Alabama Press, 1980.

Junius, R. *The Drunkards Character, Or, A True Drunkard.* London: R. Badger, 1638.

[Kempe, William]. *The Education of children in learning: Declared by the Dignitie, Utilitie, and Method thereof.* London: Thomas Orwin, 1588.

King, Henry. *A Sermon Preached at White-Hall in Lent. 1625 in Two Sermons Preached at White-Hall in Lent, March 3. 1625 and Februarie 20. 1626.* London: John Haviland, 1627.

Le Grand, Anthony. *A Dissertation Of the want of Sense and Knowledge in Brutes,* in *An Entire Body of Philosophy, According to the Principles of the Famous Renate Des Cartes.* London: Samuel Roycroft, 1694.

Le Loyer, Pierre. *A Treatise of Specters or straunge Sights, Visions and Apparitions appearing sensibly vnto men.* London: no publisher, 1605.

Lessius, Leonard. *Hygiasticon: or, the right course of preserving Life and Health unto extream old Age.* 3rd ed. Cambridge: Printers to the University, 1636.

Lever, Ralph. *The Arte of Reason, rightly termed, Witcraft, teaching a perfect way to argue and dispute.* London: Henrie Bynneman, 1573.

Lupton, Donald. *Obiectorvm Redvctio: Or, Daily Imployment for the Soule.* London: John Norton, 1634.

Maplet, John. *A Greene Forest, or a naturall Historie.* London: Henry Denham, 1567.

Markham, Gervase. *Cauelarice, Or the English Horseman.* London: Edward White, 1607.

Mascall, Leonard. *The first booke of Cattell.* London: Iohn Wolfe, 1587.

May, Edward. *Epigrams Divine and Morall.* London: I. B., 1633.

Montaigne, Michel de. *The Complete Works.* Translated by Donald M. Frame. 1943. Reprint, London: Everyman, 2003.

More, Henry, *Divine Dialogues, Containing Disquisitions Concerning the Attributes and Providence of God.* 1688. Reprint, Glasgow: Robert Foulis, 1743.

——. Correspondence. In "Descartes and Henry More on the Beast-Machine—A Translation of their Correspondence Pertaining to Animal Automatism," by Leonora D. Cohen. *Annals of Science* 1 (1936): 48–61.

Morgan, Nicholas. *The Perfection of Horse-manship, drawne from Nature; Arte, and Practise.* London: Edward White: 1609.

[Morton, Thomas]. *A Direct Answer Unto the Scandalous Exceptions which Theophilus Higgons hath lately objected against D. Morton.* London, 1609.

[Muffett, Peter]. *A Commentarie Vpon the Booke of the Prouerbes of Salamon.* London: Robert Field, 1592.

Nashe, Thomas. *Haue with you to Saffron-walden. Or, Gabriell Harueys Hunt is vp.* London: John Danter, 1596.

Nemesius. *The Natvre of Man . . . Englished, And divided into Sections, with briefs of their principall Contents: by Geo: Wither.* London: Henry Taunton, 1636.

[Nixon, Anthony]. *The Dignitie of Man, Both in the Perfections of his Soule and Bodie.* Oxford: Ioseph Barnes, 1616.

Oldenburg, Henry. *The Correspondence of Henry Oldenberg.* Translated by A. Rupert Hall and Marie Boas Hall. Madison: University of Wisconsin Press, 1965.

Ovid. *The XV. Bookes of P. Ouidius Naso, entytuled Metamorphosis, translated oute of Latin into English meeter, by Arthur Golding Gentleman*. London: William Seres, 1567.

[Peacham, Henry]. *The Mastive, or Young-Whelpe of the Olde-Dogge*. London: Tho. Creede, 1615.

Peele, George. *Merrie Conceited Iests*. London: F. Faulkner, 1627.

Perkins, William. *The Works of that Famous and Worthie Minister of Christ in the University of Cambridge, Mr William Perkins*. 3 vols. London: J. Legatt, 1616–18.

Plato, *Phaedrus*. Translated by Alexander Nehamas and Paul Woodruff. Indianapolis: Hackett, 1995.

——. *Philebus*. Translated by Harold N. Fowler. London: William Heinemann, 1962.

——. *Timaeus*. Translated by Desmond Lee. Harmondsworth: Penguin, 1971.

Pliny. *The Historie of the World. Commonly called, The Naturall Historie of C. Plinius Secvndvs*. Translated by Philemon Holland. London: A. Islip, 1601.

Plutarch. *The Philosophie, commonlie called, The Morals Written by the learned Philosopher Plutarch of Chærnea. Translated out of Greeke into English, and conferred with the Latine translations and the French, by Philemon Holland of Coventrie, Doctor in Physicke*. London: Arnold Hatfield, 1603.

Prynne, William. *Histrio-Mastix*. London: Michael Sparke, 1633.

Raleigh, Sir Walter. *The Historie of the World. In Five Bookes*. London: Walter Burre, 1614.

Rawlinson, John. *Mercy to a Beast*. Oxford: Joseph Barnes, 1612.

[Rid, Samuel]. *The Art of Jugling or Legerdemaine*. London: G. Eld, 1614.

[Rogers, Thomas]. *A philosophicall discourse, Entituled, The Anatomie of the minde*. London: Andrew Maunsell, 1576.

Sandys, Sir Miles. *Prudence, The first of the Foure Cardinall Virtues*. London: W. Sheares, 1634.

Scot, Reginald. *The discoverie of witchcraft, Wherein the lewde dealing of witches and witchmongers is notablie detected*. London: W. Brome, 1584.

Seneca. *The Workes of Lvcivs Annævs Seneca, Both Morall and Naturall*. Translated by Thomas Lodge. London: William Stansby, 1614.

Sextus Empiricus. *Outlines of Scepticism*. Translated by Julia Annas and Julian Barnes. Cambridge: Cambridge University Press, 2000.

——. *Against the Ethicists (Adversos Mathmaticos XI)*. Translated by Richard Bett. Oxford: Clarendon Press, 1997.

——. *Against the Grammarians (Adversos mathematicos I)*. Translated by D. L. Blank. Oxford: Clarendon Press, 1998.

Shakespeare, William. *The Complete Works*. Edited by Stanley Wells and Gary Taylor. Oxford: Oxford University Press, 1988.

Spackman, Thomas. *A Declaration of Svch Greivovs accidents as commonly follow the biting of mad Dogges, together with the cure thereof*. London: John Bill, 1613.

Spenser, Edmund. *The Faerie Queene*. In *The Poetical Works of Edmund Spenser*, edited by J. C. Smith and E. De Selincourt. Oxford: Oxford University Press, 1924.

Stowe, John, and Edmond Howes. *The Annales, Or Generall Chronicle of England*. London: Thomæ Adams, 1615.

Strode, William. *The Floating Island: A Tragicomedy, Acted before his Majesty at OX-FORD, Aug.29.1636. By the Students of CHRIST-CHURCH.* London: H. Twiford, 1655.

Stubbes, Phillip. *The Anatomie of Abuses.* London: J. R. Jones, 1583.

Tarlton, Richard. *Tarltons Jests.* London: A. Crook, 1638.

Taylor, John. *The Nipping and Snipping of Abuses.* London: Ed. Griffin, 1614.

Thompson, Thomas. *A Diet for a Drvnkard.* London: Richard Bankworth, 1612.

Topsell, Edward. *The Historie of Fovre-Footed Beastes.* London: William Jaggard, 1607.

———. *The Historie of Serpents. Or, The second Booke of liuing Creatures.* London: William Jaggard, 1608.

———. *The Fowles of Heauen or History of Birdes.* c. 1613–14. Edited by Thomas P. Harrison and F. F. David Hoeniger. Austin: University of Texas Press, 1972.

Turbervile, George. *The Noble Arte of Venerie or Hvnting.* London: Henry Bynneman, 1575.

———. *The Booke of Faulconrie or Hauking, For the Onely Delight and pleasure of all Noblemen and Gentlemen.* London: Christopher Barker, 1575.

Tylney, Robert. *Two Learned Sermons.* London: W. Hall, 1609.

Vicary, Thomas. *A profitable Treatise of the Anatomie of mans body.* 1548. Reprint, London: Henry Bamforde, 1577.

Viret, Pierre. *The Schoole of Beastes, Intituled, the good Housholder, or the Oeconomickes.* Translated by I. B. London: Robert Wal-de-graue, 1585.

Vives, Juan Luis. *The Passions of the Soul: The Third Part of De Anima et Vita.* Translated by Carlos G. Noreña. Lewiston, Queenston, Lampeter: The Edwin Mellen Press, 1990.

[Walkington, Thomas]. *The Optick Glasse of Hvmors.* London: John Windet, 1607.

Whitaker, Alexander. *Good Newes From Virginia.* London: F. Kyngston, 1613.

White, T. H., ed. *The Book of Beasts: Being a Translation from a Latin Bestiary of the Twelfth Century.* Stroud: Alan Sutton, 1992.

Whitney, Geffrey. *A Choice of Emblemes.* Leyden: Christopher Plantyn, 1586.

Widdowes, Daniel. *Natural Philosophy: Or A Description of the World, and of the severall Creatures therein contained.* 2nd ed. London: T. Cotes, 1631.

[Wilcox, Thomas]. *A Short, Yet sound Commentarie; written on that woorthie worke called; The Prouerbes of Salomon.* London: Thomas Orwin, 1589.

Wilkinson, Robert. *The Stripping of Joseph, Or the crueltie of Brethren to a Brother.* London: Henry Holland, 1625.

Willis, Thomas. *Two Discourses Concerning the Soul of Brutes, Which is that of the Vital and Sensitive of Man.* London: Thomas Dring, 1683.

Wilson, George. *The Commendation of Cockes, and Cock-Fighting.* London: Henrie Tomes, 1607.

Woolton, John. *A Newe Anatomie of whole man, aswell of his body, as of his Soule.* London: Thomas Purfoote, 1576.

———. *A Treatise of the Immortalitie of the Soule.* London: John Shepperd, 1576.

[Wright, Thomas]. *The Passions of the Minde.* London: V. S. for W. B., 1601.

Young, Thomas. *Englands Bane: Or, The Description of Drunkennesse.* London: William Jones, 1617.

Note on Secondary Sources

The following is a selective list of secondary materials mentioned in this book or otherwise relevant to the issues raised in each chapter.

General

The history of animals is an emerging area of scholarship, and an overview of recent work and some of the ethical concerns is in my "A Left-Handed Blow: Writing the History of Animals," in *Representing Animals,* ed. Nigel Rothfels (Bloomington: Indiana University Press, 2002), 3–18. Important among the earlier work in the history of animals are Dix Harwood, *Love for Animals and How It Developed in Great Britain* (1928; repr., New York: Edwin Mellen Press, 2002) and E. S. Turner, *All Heaven in a Rage* (1964; repr., Fontwell, Sussex: Centaur, 1992). More recent studies include Kathleen Kete, *The Beast in the Boudoir: Petkeeping in Nineteenth-Century Paris* (Berkeley, Los Angeles, and London: University of California Press, 1994); Harriet Ritvo, *The Animal Estate: The English and Other Creatures in the Victorian Age* (London: Penguin, 1990); Ritvo, *The Platypus and the Mermaid and Other Figments of the Classifying Imagination* (Cambridge, Mass.: Harvard University Press, 1997); Louise E. Robbins, *Elephant Slaves and Pampered Pets: Exotic Animals in Eighteenth-Century Paris* (Baltimore: Johns Hopkins University Press, 2002); Nigel Rothfels, *Savages and Beasts: The Birth of the Modern Zoo* (Baltimore: Johns Hopkins University Press, 2002); and James Serpell, *In the Company of Animals: A Study of Human-Animal Relationships* (Oxford: Blackwell, 1986).

Studies of animals in premodern cultures include Richard Sorabji's important *Animal Minds and Human Morals: The Origin of the Western Debate* (London: Duckworth, 1993). This remains the most wide-ranging study of classical ideas about reason and animals, while Roger French's *Ancient Natural History: Histories of Nature* (London: Routledge, 1994) offers an overview of that field. Joyce E. Salisbury's *The Beast Within: Animals in the Middle Ages* (London: Routledge, 1994) and Dorothy Yamamoto's *The Boundaries of the Human in Medieval English Literature* (Oxford: Oxford University Press, 2000) are the key studies of medieval ideas, while Salisbury's collection of essays, *The Medieval World of Nature: A Book of Essays* (New York: Garland, 1993) offers further work, as do Nona C. Flores, ed., *Animals in the Middle Ages* (New York: Routledge, 1996) and Debra Hassig, ed., *The Mark of the Beast* (New York: Routledge, 1999).

Focusing on animals in early modern society are Bruce Boehrer, *Shakespeare Among the Animals: Nature and Society in the Drama of Early Modern England* (Basingstoke: Palgrave, 2002); Erica Fudge, *Perceiving Animals: Humans and Beasts in Early Modern English Culture* (Basingstoke: Palgrave, 2000); Keith Thomas, *Man and the Natural World: Changing Attitudes in England 1500–1800* (1983; repr., London: Penguin, 1994). Other works include Erica Fudge, ed., *Renaissance Beasts: Of Animals, Humans. and Other Wonderful Creatures* (Urbana: University of Illinois Press, 2004). My introduction to this collection (pp. 1–17) contains an overview of recent essays and articles on animals in early modern culture. Peter Harrison has published widely on animals in seventeenth-century thought, and some of his most important essays are included under the relevant chapter headings below. Two essays on the use of dogs in humanist iconography are Patrik Reutersward, "The Dog in the Humanist's Study," in *The Visible and Invisible in Art: Essays in the History of Art* (Vienna: IRSA, 1991), 206–25, and Karl Josef Höltgen, "Clever Dogs and Nimble Spaniels: On the Iconography of Logic, Invention, and Imagination," *Explorations in Renaissance Culture* 24 (1998): 1–36.

I am also indebted to work coming from broader fields of inquiry, and would include in particular here Jacques Derrida's essay "The Animal That Therefore I Am (More to Follow)," trans. David Wills, *Critical Inquiry* 28 (2002): 369–418.

Chapter 1. Being Human

Jonathan Sawday's *The Body Emblazoned: Dissection and the Human Body in Renaissance Culture* (London: Routledge, 1995) and Michael C. Schoenfeldt, *Bodies and Selves in Early Modern England: Physiology and Inwardness in Spenser, Shakespeare, Herbert and Milton* (Cambridge: Cambridge University Press,

1999) offer different assessments of the cultural impact of developments in science and medicine in the period. The essays in "Part I: Renaissance and Early Modern," (pp. 17–57) of Roy Porter, ed., *Rewriting the Self: Histories from the Renaissance to the Present* (London: Routledge, 1997) offer overviews of arguments from Petrarch to Descartes.

Useful introductions to Aristotle's ideas are Jonathan Barnes, *Aristotle* (Oxford: Oxford University Press, 1982) and Stephen Everson, "Psychology," in *The Cambridge Companion to Aristotle*, ed. Jonathan Barnes (Cambridge: Cambridge University Press, 1995), 168–94. Lynn S. Joy's essay "Epicureanism in Renaissance Moral and Natural Philosophy," *Journal of the History of Ideas* 53, no. 4 (1992): 573–83, provides an important context, as does Gillian Clark's "The Fathers and the Animals: The Rule of Reason?," in *Animals on the Agenda: Questions about Animals for Theology and Ethics*, ed. Andrew Linzey and Dorothy Yamamoto (Urbana: University of Illinois Press, 1998), 67–79, which focuses on St. Augustine.

Key studies of the influence of Aristotle on early modern conceptions of the self include Simon Kemp, *Medieval Psychology* (Contributions in Psychology 14, New York: Greenwood Press, 1990); Katherine Park and Eckhard Kessler, "The Concept of Psychology," (455–63), Katherine Park, "The Organic Soul," (464–84), and Eckhard Kessler, "The Intellective Soul," (485–534) in *The Cambridge Companion to Renaissance Philosophy*, ed. Charles B. Schmitt and Quentin Skinner (Cambridge: Cambridge University Press, 1988); Nancy G. Siraisi, *Medieval and Early Renaissance Medicine: An Introduction to Knowledge and Practice* (Chicago and London: University of Chicago Press); Nicholas H. Steneck, "Albert the Great on the Classification and Localization of the Internal Senses," *Isis* 65 (1975): 193–211; and David Summers, *The Judgment of Sense: Renaissance Naturalism and the Rise of Aesthetics* (Cambridge University Press, 1987). On early modern ideas see Ruth Leila Anderson, "Elizabethan Psychology and Shakespeare's Plays," *University of Iowa Humanistic Studies* 3.4 (1927); Lawrence Babb, *The Elizabethan Malady: A Study in Melancholia in English Literature from 1580 to 1643* (East Lansing: Michigan State College Press, 1951); J. B. Bamborough, *The Little World of Man* (London: Longman, 1952); Murray W. Bundy, "Shakespeare and Elizabethan Psychology," *Journal of English and Germanic Philology* 23 (1924): 516–49; E. Ruth Harvey, *The Inward Wits: Psychological Theory in the Middle Ages and the Renaissance* (London: Warburg Institute, 1975). These older studies have yet to be replaced by more recent work in the area, but essays in Neil Rhodes and Jonathan Sawday, eds., *The Renaissance Computer: Knowledge Technology in the First Age of Print* (London and New York: Routledge, 2000) offer new assessments of mind and memory, while Peter Harrison's "Original Sin and the Problem of Knowledge in Early Modern Europe," *Journal of the History of Ideas* 63.2 (2002): 239–59, is also helpful.

Stephen Gaukroger's collection *The Soft Underbelly of Reason: The Passions in the Seventeenth Century* (London and New York: Routledge, 1998) contains another important essay by Peter Harrison, "Reading the Passions: The Fall, the passions, and dominion over nature," (pp. 49–78), as well as Susan James's "Explaining the Passions: Passions, desires, and the explanation of action," (pp. 17–33). An older study of the role of the passions in the construction of the human is Lily B. Campbell, *Shakespeare's Tragic Heroes: Slaves of Passion* (1930; repr., London: Methuen, 1961), while Gillian Clark's "Animal Passions," *Greece and Rome* 47.1 (2000): 88–93, focuses on the passions and the reasoning capacities of animals. More recently Gail Kern Paster's *Humoring The Body: Emotions and the Shakespearean Stage* (Chicago: University of Chicago Press, 2004) focuses in one chapter, 135–188, on the place of animals in discussions of early modern emotions.

Irven M. Resnick and Kenneth F. Kitchell Jr.'s "Albert The Great on the 'Language' of Animals," *American Catholic Philosophical Quarterly* 70.1 (1996): 41–61, provides an important overview of classical ideas about animal language, while R. W. Serjeantson's "The Passions and Animal Language, 1540–1700," *Journal of the History of Ideas* 62.3 (2001): 425–44, offers an assessment of early modern arguments. Brian Cummings's essay, "Pliny's Literate Elephant and the Idea of Animal Language in Renaissance Thought," in *Renaissance Beasts,* ed. Fudge, 164–85, links debates about animal language with skepticism, while Matthew Senior's "'When the Beasts Spoke': Animal Speech and Classical Reason in Descartes and La Fontaine," in *Animal Acts: Configuring the Human in Western History,* ed. Jennifer Ham and Matthew Senior (London and New York: Routledge, 1997), 61–84, links the scientific revolution with the silencing of animals.

On laughter, key sources include Helen Adolf, "On Medieval Laughter," *Speculum* 22 (1947): 251–53; R. E. Ewin, "Hobbes on Laughter," *The Philosophical Quarterly* 51.202 (2001): 29–40; Stephen Halliwell, "The Uses of Laughter in Greek Culture," *Classical Quarterly* 41.2 (1991): 279–96; M. A. Screech and Ruth Calder, "Some Renaissance Attitudes to Laughter," in *Humanism in France at the End of the Middle Ages and in the Early Renaissance,* ed. A. H. T. Levi (Manchester: Manchester University Press, 1970), 216–28; M. A. Screech, *Laughter at the Foot of the Cross* (London: Penguin, 1997); and Quentin Skinner, "Why Laughing Mattered in the Renaissance: The Second Henry Tudor Memorial Lecture," *History of Political Thought* 22.3 (2001): 418–47. My own essay, "Learning to Laugh: Children and being human in early modern thought," *Textual Practice* 17.2 (2003): 277–94, traces the use of laughter in the education and training of children in early modern England.

The chapter on prudence in Joseph Pieper's *The Four Cardinal Virtues: Prudence, Justice, Fortitude, Temperance* (Notre Dame: University of Notre

Dame Press, 1966) offers a useful overview of that virtue, as does Lynette C. Black's "Prudence in Book II of *The Faerie Queene*," *Spenser Studies* 13 (1999): 65–90. Robert Hoope's *Right Reason in the English Renaissance* (Cambridge, Mass.: Harvard University Press, 1962) remains an important assessment of reason and prudence in the period. Victoria Kahn's *Rhetoric, Prudence, and Skepticism in the Renaissance* (Ithaca and London: Cornell University Press, 1985) provides a different context in which to place prudence. Mark A. Holowchak's "Aristotle on Dreaming: What Goes On in Sleep When the 'Big Fire' Goes Out," *Ancient Philosophy* 16.2 (1996): 405–23, offers an important overview. Steven F. Kruger's *Dreaming in the Middle Ages* (Cambridge: Cambridge University Press, 1992) is an exhaustive study of classical and medieval dream theorists and provides an important background to early modern ideas. A. Roger Ekirch's essay, "Sleep We Have Lost: Pre-Industrial Slumber in the British Isles," *The American Historical Review* 106.2 (2001): 343–86, outlines the sleep patterns of early modern men and women and shows how this distinct nature of sleep (consisting of a "first sleep" followed by a period of wakefulness and then a "second sleep") may have offered a different context into which historians should place early modern dreams. More general studies of early modern dreams include Peter Brown ed., *Reading Dreams: The Interpretation of Dreams from Chaucer to Shakespeare* (Oxford: Oxford University Press, 1999); Peter Burke, "The Cultural History of Dreams," in his *Varieties of Cultural History* (Cambridge: Polity Press, 1997), 23–42; Patricia Crawford, "Women's Dreams in Early Modern England," *History Workshop Journal* 49 (2000): 129–41; Marjorie B. Garber, *Dream in Shakespeare: From Metaphor to Metamorphosis* (New Haven and London: Yale University Press, 1974); S. R. F. Price, "The Future of Dreams: From Freud to Artemidorus," *Past and Present* 113 (1986): 3–37; and Janine Rivière, "'Visions of the Night': The Reform of Popular Dream Beliefs in Early Modern England," *Parergon* 20.1 (2003): 109–38. Alan Macfarlane offers an analysis of some of the dreams experienced by Ralph Josselin in the mid-seventeenth century in *The Family Life of Ralph Josselin a Seventeenth-Century Clergyman: An Essay in Historical Anthropology* (Cambridge: Cambridge University Press, 1970), 183–87.

Chapter 2. Becoming Human

Debates about the nature of the role of women and men in reproduction are outlined in Anthony Preus, "Galen's Criticism of Aristotle's Conception Theory," *Journal of the History of Biology* 10.1 (1977): 65–85, and Michael Boylan, "The Galenic and Hippocratic Challenges to Aristotle's Conception Theory," *Journal of the History of Biology* 17.1 (1984): 83–112. Thomas Laqueur's *Making Sex: Body and Gender from the Greeks to Freud* (Cambridge,

Mass.: Harvard University Press, 1990) looks, in chapter 3 (pp. 63–113), at Renaissance conceptions of sex and reproduction. For a historical overview of ideas about the embryo see G. R. Dunstan, "The Human Embryo in the Western Moral Tradition," in *The Status of the Human Embryo: Perspectives from Moral Tradition*, ed. G. R. Dunstan and Mary J. Sellar (Oxford: Oxford University Press, 1988), 39–57. Dunstan's collection, *The Human Embryo: Aristotle and the Arabic and European Traditions* (Exeter: University of Exeter Press, 1990), contains useful essays including Helen King's "Making a man: becoming human in early Greek medicine" (pp. 10–19) and Pamela M. Huby's "Soul, Life, Sense, Intellect: Some Thirteenth-Century Problems" (pp. 113–22). Peter J. Bowler's "Preformation and Pre-existence in the Seventeenth Century: A Brief Analysis," *Journal of the History of Biology* 4.2 (1971): 221–44, offers an overview of debates surrounding the "belief in the existence of a miniature organism at some time before conception" that were taking place in the second half of the seventeenth century; and Matthew R. Goodrum's "Atomism, Atheism and the Spontaneous Generation of Human Beings: The Debate over a Natural Origin of the First Humans in Seventeenth-Century Britain," *Journal of the History of Ideas* 63.2 (2002): 207–25, looks at natural theories of the origin of humanity.

There are numerous studies of early modern attitudes toward, and treatment of, children, and these include: Patrick Collinson, "The Protestant Family," in his *Birthpangs of Protestant England: Religion and Cultural Change in the 16th and 17th Centuries* (Basingstoke: Macmillan, 1988), 60–93; Ian Green, "'For Children in Yeeres and Children in Understanding': The Emergence of the English Catechism under Elizabeth and the Early Stuarts," *Journal of Ecclesiastical History* 37.3 (1986): 397–425; Philip Greven, *The Protestant Temperament: Patterns of Child Rearing, Religious Experience and the Self in Early America* (New York: Meridian, 1979); Christopher Hill, "The Spiritualization of the Household," in his *Society and Puritanism in Pre-Revolutionary England* (1964; repr., Harmondsworth: Penguin, 1986), 429–66; John Morgan, *Godly Learning: Puritan Attitudes Towards Reason, Learning and Education, 1560–1640* (Cambridge: Cambridge University Press, 1986); Robert V. Schnucker, "Puritan Attitudes Towards Childhood Discipline, 1560–1634," in *Women as Mothers in Pre-Industrial England: Essays in Memory of Dorothy McLaren*, ed. Valerie Fildes (London and New York: Routledge, 1990), 108–21; J. A. Sharpe, "Disruption of the Well-Ordered Household: Age, Authority and Possessed Young People," in *The Experience of Authority in Early Modern England*, ed. Paul Griffiths, Adam Fox, and Steve Hindle (Basingstoke: Macmillan, 1996), 187–212; C. John Sommerville, *The Discovery of Childhood in Puritan England* (Athens and London: University of Georgia Press, 1992); Keith Thomas, "Age and Authority in Early Modern England," *Proceedings of the British Academy* 62 (1976): 205–48; and Keith Thomas, "Children in Early Modern England," in *Children and their Books: A Celebration of the Work of Iona and Peter Opie*, ed. Gillian Avery and Julia Briggs (Oxford: Oxford University Press, 1989), 45–77. Helen M. Jewell's *Education in Early Modern England* (Basingstoke: Macmillan, 1998) offers an overview of that institution. Anthony Fletcher has argued that in early modern ideas, masculinity was not given at birth but had to be attained; my argument about the attainment of humanity echoes his ideas but takes them into the realm of species rather than gender; see Fletcher, "Manhood, the male body, courtship and the household in early modern England," *History* 84.275 (1999): 419–36. Norbert Elias's *The Civilizing Process* (Oxford: Blackwell, 1994) remains the classic study of the development of "self-control" from the medieval and Renaissance worlds.

Margaret T. Hodgen's *Early Anthropology in the Sixteenth and Seventeenth Century* (1964; repr., Philadelphia: University of Pennsylvania Press, 1971) offers an important overview of the ideas and challenges faced by early modern travelers. More recent studies of colonial activities include Peter Hulme, *Colonial Encounters: Europe and the Native Caribbean 1492–1797* (London: Methuen, 1986); Anthony Pagden, *The Fall of Natural Man: The American Indian and the Origins of Comparative Ethnology* (Cambridge: Cambridge University Press, 1982); and Olive P. Dickason, *The Myth of the Savage and the Beginnings of French Colonialism in the Americas* (1984; repr., Edmonton, Can.: University of Alberta Press, 1997). On the work of the Virginia Company, John Parker argues for the primary importance of religion in the establishment of the Virginia colony, while Wesley Frank Craven and T. H. Breen make alternative arguments in favor of the priority of the economic. Andrew Fitzmaurice proposes that the ideology of the Virginia Company is civic and humanist. See Parker, "Religion and the Virginia Company, 1609–10," in *The Westward Enterprise: English Activities in Ireland, the Atlantic and America 1480–1650*, ed. K. R. Andrews, N. P. Canny, and P. E. H. Hair (Liverpool: Liverpool University Press, 1978), 245–70; Craven, *Dissolution of the Virginia Company: The Failure of a Colonial Experiment* (New York: Oxford University Press, 1932), esp. p. 24; Breen, *Puritans and Adventurers: Change and Persistence in Early America* (Oxford: Oxford University Press, 1980), esp. pp. 106–26; Fitzmaurice, "The Civic Solution to the Crisis of English Colonization, 1609–1625," *The Historical Journal* 42.1 (1999): 25–51; and Fitzmaurice, "Classical Rhetoric and the Promotion of the New World," *Journal of the History of Ideas* 58.2 (1997): 221–43. Brian Cummings's study of blushing reveals another way in which the natives of the New World were felt to be less than human: Cummings, "Animal Passions and Human Sciences: Shame, Blushing and Nakedness in Early Modern England and the New World," in *At the Borders of the Human: Beasts, Bodies and Natural Philosophy in the Early Modern Period*, ed. Erica Fudge, Ruth Gilbert, and Susan Wiseman (Basingstoke: Macmillan, 1999), 26–50. I read Virginia DeJohn Anderson's *Creatures of Empire: How Domestic Animals Transformed Early America* (Oxford: Oxford University Press, 2004) too late to fully integrate her important argument about the role cattle farming played in establishing the difference between native and English settler populations in the New World, but it is clearly relevant to my discussions.

Chapter 3. Becoming Animal

Many of the works listed under chapter 1 offer important outlines of the sense of the fragility of human status, but without fully articulating the role of animals. Nancy E. Snow's essay, "Compassion," *American Philosophical Quarterly* 28.3 (1991): 195–205, offers a suggestive analysis of the subject. On Seneca, see Daniel Baraz, "Seneca, Ethics and the Body: The Treatment of Cruelty in Medieval Thought," *Journal of the History of Ideas* 59.2 (1998): 195–215. On Aquinas, see Judith Barad, "Aquinas' Inconsistency on the Nature and the Treatment of Animals," *Between the Species* 88.4 (1988): 102–11; Peter Drum, "Aquinas and the Moral Status of Animals," *American Catholic Philosophical Quarterly* 66.4 (1992): 483–88; Stephen Loughlin, "Similarities and Differences Between Human and Animal Emotion in Aquinas's Thought," *The Thomist* 65 (2001): 45–65; and Dorothy Yamamoto, "Animals and Aquinas: Patrolling the Boundary?" in *Animals on the Agenda,* ed. Linzey and Yamamoto, 80–89. The classic study of the organization of the natural world is Arthur O. Lovejoy, *The Great Chain of Being: A Study of the History of an Idea* (1936; repr., Cambridge, Mass. and London: Harvard University Press, 1976).

Edward Berry notes the link between Montaigne and *As You Like It* but offers a very different reading in "Pastoral Hunting in *As You Like It,*" in his *Shakespeare and the Hunt: A Cultural and Social Study* (Cambridge: Cambridge University Press, 2001), 159–89. In *Ordinary Vices* (Cambridge, Mass. and London: Bellknap Press, 1984) Judith N. Shklar looks at the place of cruelty in philosophy and reads Montaigne as a key philosopher of cruelty. Of the numerous works on Montaigne, those that have been most helpful to me include Philip P. Hallie, "The Ethics of Montaigne's 'De la Cruauté,'" in *O Un Amy! Essays on Montaigne in Honor of Donald M. Frame,* ed. Raymond C. La Charité (Lexington, Ky.: French Forum, 1977), 156–71. David Quint's essay "Letting Oneself Go: 'Of Anger' and Montaigne's Ethical Reflections," *Philosophy and Literature* 24:1 (2000) and his book *Montaigne and the Quality of Mercy: Ethical and Political Ideas in the Essais* (Princeton, Nj.: Princeton University Press, 1998): both concentrate on Montaigne's ethics. Dan Engster's

"The Montaignian Moment," *Journal of the History of Ideas* 594 (1998): 625–50 traces the role of nature in Montaigne's thought.

On Montaigne in English culture, a number of essays concentrate on Shakespeare's knowledge of Montaigne, and in doing so offer useful overviews of the *Essais* in England. See, for example, Robert Ellrodt, "Self-Consciousness in Montaigne and Shakespeare," *Shakespeare Survey* 28 (1975). 37–50; Fred Parker, "Shakespeare's Argument with Montaigne," *Cambridge Quarterly* 28.1 (1999): 1–18. Katherine Kerestman reads Montaigne's true impact on attitudes toward animals as being felt in the eighteenth century in "Breaking the Shackles of the Great Chain of Being and Liberating Compassion in the Eighteenth Century," in *1650–1850: Ideas, Aesthetics, and Inquiries in the Early Modern Era,* ed. Kevin L. Cope (New York: AMS Press, 1997), 3:57–76.

Chapter 4. Being Animal

On Plutarch's influence on assessments of human and animal behavior, see Robert J. Richards, "Influence of Sensationalist Tradition on early Theories of the Evolution of Behavior," *Journal of the History of Ideas* 40.1 (1979): 85–105. On Spenser's rendition of the Platonic human see Jerry Leath Mills, "Spenser, Lodowick Bryskett and the Mortalist Controversy: *The Faerie Queene,* II, ix, 22," *Philological Quarterly* 52.2 (1973): 173–86; Robert L. Reid, "Alma's Castle and the Symbolization of Reason in *The Faerie Queene,*" *Journal of English and Germanic Philology* 80.4 (1981): 512–27; Reid, "Spenserian Psychology and the Structures of Allegory in Books 1 and 2 of *The Faerie Queene,*" *Modern Philology* 79.4 (1982): 359–75; Schoenfeldt, *Bodies and Selves,* 40–73.

George Boas's classic study *The Happy Beast in French Thought of the Seventeenth Century* (1933; repr., New York: Octagon Books, 1966) reads Montaigne and Charron as theriophiles. An assessment of Montaigne's skepticism is in Donald M. Frame, *Montaigne's Discovery of Man: The Humanization of a Humanist* (New York: Columbia University Press, 1955). More recent considerations include Floyd Gray, "Montaigne's Pyrrhonism," in *O Un Amy!,* ed. La Charité, 119–36; William M. Hamlin, "On Continuities Between Skepticism and Early Ethnography; Or, Montaigne's Providential Diversity," *Sixteenth Century Journal* 31.2 (2000): 361–79; Zachary S. Schiffman, "Montaigne and the Rise of Skepticism in Early Modern Europe: A Reappraisal," *Journal of the History of Ideas* 45.4 (1984): 499–516; and David L. Sedley, "Sublimity and Skepticism in Montaigne," *PMLA* 113.5 (1998): 1079–92. Peter Harrison's "The Virtues of Animals in Seventeenth-Century Thought," *Journal of the His-*

tory of Ideas 59.3 (1998): 463–84 offers an overview of some of the key ways of reading animals in the period, and follows Boas in assessing Montaigne as a theriophile. Tullio Gregory offers an assessment of Charron in "Pierre Charon's 'Scandalous Book,'" (pp. 87–109), and Nigel Smith looks at the impact of Montaigne on writings of the English Civil War in "The Charge of Atheism and the Language of Radical Speculation, 1640–1660," (pp. 131–58): both are in *Atheism from the Reformation to the Enlightenment,* ed. Michael Hunter and David Wootton (Oxford: Clarendon, 1992). Alan Stewart's essay "Government by Beagle: The Impersonal Rule of James VI and I," 101–15, in *Renaissance Beasts,* ed. Fudge, offers an analysis of the political implications of James's love of hunting.

Classic analyses of the study of animals in the early modern period are Charles E. Raven, *English Naturalists from Neckam to Ray: A Study in the Making of the Modern World* (Cambridge: Cambridge University Press, 1947) and Allen Debus, *Man and Nature in the Renaissance* (Cambridge: Cambridge University Press, 1978). Peter Harrison's *The Bible, Protestantism, and the Rise of Natural Science* (Cambridge: Cambridge University Press, 1998) provides an important intellectual context for reading animals in natural philosophy during the period. William B. Ashworth Jr. has traced the changing view of animals in early modern natural philosophy in two essays: "Emblematic Natural History of the Renaissance," in *Cultures of Natural History,* ed. N. Jardine, J. A. Secord, and E. Spary (Cambridge: Cambridge University Press, 1996) and "Natural History and the Emblematic World View," in *Reappraisals of the Scientific Revolution,* ed. David C. Lindberg and Robert S. Westman (Cambridge: Cambridge University Press, 1990), 303–32. Eric Baratay and Elisabeth Hardouin-Fugier's *Zoo: A History of Zoological Gardens in the West,* trans. Oliver Welsh (1998; repr., London: Reaktion, 2002) tells of the growing interest in and captivity of animals from premodern to modern times.

Jason Scott-Warren's essay "When Theatres were Bear-Gardens: or, What's at Stake in the Comedy of Humours," *Shakespeare Quarterly* 54.1 (2003): 63–82, offers an analysis of baiting in the period, while Barbara Ravelhofer looks at the importation of white bears and at their training in "'Beasts of Recreation': Henslowe's White Bears," *English Literary History* 32 (2002): 287–323. Louise Hill Curth's two essays "English Almanac and Animal Health Care in the Seventeenth Century," *Society and Animals* 8.1 (2000): 71–86, and "The Care of the Brute Beast: Animals and the Seventeenth-Century Medical Market-Place," *Social History of Medicine* 15.3 (2002): 375–92, examine early veterinary texts in the period. Claudia Lazzaro's "Animals as Cultural Signs: A Medici Menagerie in the Grotto at Castello," in *Reframing the Renaissance: Visual Culture in Europe and Latin America 1450–1650,* ed. Claire Farago (New Haven and London: Yale University Press,

1995), 197–227, links the visual representation of animals to developments in naturalism.

On classical skepticism see Jonathan Barnes, "The Beliefs of a Pyrrhonist," *Proceedings of the Cambridge Philosophical Society* 208, n.s. 28 (1982): 1–29. On early modern skepticism, the key study is Richard H. Popkin, *The History of Scepticism from Erasmus to Spinoza* (1960; repr., Berkeley and London: University of California Press, 1979). See also Luciano Floridi, "Scepticism and Animal Rationality: The Fortune of Chrysippus' Dog in the History of Western Thought," *Archiv für Geschichte der Philosophie* 79.1 (1997): 27–57; and Floridi, "The Diffusion of Sextus Empiricus's Works in the Renaissance," *Journal of the History of Ideas* 56.1 (1995): 63–85; John Christian Laursen, *The Politics of Skepticism in the Ancients, Montaigne, Hume and Kant* (New York and Cologne: E. J. Brill, 1992); José R. Maia, "Academic Skepticism in Early Modern Philosophy," *Journal of the History of Ideas* 58.2 (1997): 199–220; and Charles B. Schmitt, *Cicero Scepticus: A Study of the Influence of the Academica in the Renaissance* (The Hague: Martinus Nijhoff, 1972). William M. Hamlin's "A Lost Translation Found? An Edition of *The Sceptick* (c. 1590) Based on Extand Manuscripts [With Text]," *English Literary Renaissance* 12.2 (2001): 34–51, traces the transmission of *The Sceptick* and includes a complete edition of that text. Katherine Eisaman Maus includes a brief discussion of the role of scepticism in conceptualizations of human and animal interiority in *Inwardness and Theater in the English Renaissance* (Chicago: University of Chicago Press, 1995).

Chapter 5. A Reasonable Animal?

The two substantial studies of Morocco are J. O. Halliwell-Phillips, "The Dancing Horse," in *Memoranda on Love's Labour's Lost, King John, Othello, and on Romeo and Juliet* (London: James Evan Adlard, 1879), 21–57, and the entry "Banks" in the *D.N.B.*, 3:125–26. There is a note by Sidney H. Atkins on "Mr. Banks and his Horse" in *Notes and Queries* (21 July, 1934): 39–44, and Arthur Freeman includes a chapter on "Banks and Morocco" in his *Elizabethan Misfits: Brief Lives of English Eccentrics, Exploiters, Rogues and Failures 1580–1660* (New York: Garland, 1978), 123–39. There is also a brief discussion of the duo in Ruth Manning Sanders, *The English Circus* (London: Werner Laurie, 1952), 23–25. Thomas Sebeok looks at the horse and at Samuel Rid's arguments in relation to the early twentieth-century tale of Clever Hans, another intelligent horse, in *The Sign and Its Masters* (Austin and London: University of Texas Press, 1979), 85–106. Otherwise, Morocco seems to have been of little interest to modern scholars. Other studies of performing animals in early modern England include the classic essay by Louis B. Wright, "Animal Actors on the English Stage Before 1642," *PMLA* 42 (1927): 656–69, and more recently Matthew Bliss, "Property or Performers? Animals on the Elizabethan Stage," *Theatre Studies* (1994): 45–58, and Michael Dobson, "A Dog at All Things: The Transformation of the On-stage Canine, 1550–1850," *Performance Research* 5.2 (2000): 116–24. It is the absence of animals on the English stage that is the focus of Harry Levin's "Falstaff Uncolted," in *Shakespeare and the Revolution of the Times: Perspectives and Commentaries* (New York: Oxford University Press, 1976), 121–30.

On magic, Stuart Clark's *Thinking With Demons: The Idea of Witchcraft in Early Modern Europe* (Oxford: Oxford University Press, 1997) is the most significant modern study, while Sydney Anglo's "Reginald Scot's *Discoverie of Witchcraft:* Scepticism and Saduceeism," in his *The Damned Art: Essays in the Literature of Witchcraft*, (London: Routledge, 1977), 106–39, contextualizes that text. On Circe, see Gareth Roberts, "The descendants of Circe: witches and Renaissance fictions," in *Witchcraft in Early Modern Europe: Studies in Culture and Belief*, ed. Jonathan Barry, Marianne Hester, and Gareth Roberts (Cambridge: Cambridge University Press, 1996), 183–206; and Marina Warner, "The Enchantments of Circe," *Raritan* 17 (1997): 1–23.

Peter Sobol offers an important assessment of medieval ideas about animal intelligence in "The Shadow of Reason: Explanations of Intelligent Animal Behaviour in the Thirteenth Century," in *The Medieval World of Nature*, ed. Salisbury, 109–28. On horse training, and especially Gervase Markham, see Elspeth Graham's essay "Reading, Writing, and Riding Horses in Early Modern England: James Shirley's *Hyde Park* (1632) and Gervase Markham's *Cavelarice* (1607)," in *Renaissance Beasts*, ed. Fudge, 116–37.

Chapter 6. A Reasonable Human?

Stephen Gaukroger's *Descartes: An Intellectual Biography* (Oxford: Oxford University Press, 1995) is an important interpretation of the thinker in his context, and the essays in *The Cambridge Companion to Descartes*, ed. John Cottingham (Cambridge: Cambridge University Press, 1992) offer an overview of his philosophy. Leonora Cohen-Rosenfield's *From Beast-Machine to Man-Machine: Animal Soul in French Letters from Descartes to La Mettrie* (1940; repr., New York: Octagon Books, 1968) provides an outline of Descartes's influence, particularly in France.

There is a vast body of scholarly material on René Descartes, and what follows here concentrates particularly on the beast-machine hypothesis. Key essays are John Cottingham, "A 'Brute To Brutes'?: Descartes' Treatment of Animals," *Philosophy* 53 (1978): 551–59; Peter Harrison, "Descartes on Animals," *Philosophical Quarterly* 42 (1992): 219–27; Peter Harrison, "Theodicy and Animal Pain," *Philosophy* 64 (1989): 79–92; Peter Harrison, "Do Animals Feel Pain?," *Philosophy* 66.1 (1991): 25–40; Katherine Morris, "*Bêtes-Machines*," in *Descartes' Natural Philosophy,* ed. Stephen Gaukroger, John Schuster, and John Sutton (London and New York: Routledge, 2000), 401–19; Lex Newman, "Unmasking Descartes's Case for the *Bête Machine* Doctrine," *Canadian Journal of Philosophy* 31.3 (2001): 389–425; Stefan Sencerz, "Descartes on Sensations and 'Animal Minds,'" *Philosophical Papers* 19.2 (1990): 119–41; and Gary Steiner, "Descartes on the Moral Status of Animals," *Archiv für Geschichte der Philosophie* 80.3 (1998): 268–91. Michael Miller attempts to challenge the beast-machine hypothesis, using empirical evidence from modern experiments with chimpanzees and their acquisition of sign language: Miller, "Descartes' Distinction Between Animals and Humans: Challenging the Language and Action Tests," *American Catholic Philosophical Quarterly* 72.3 (1998): 339–70. Peter Harrison's "Animal Souls, Metempsychosis and Theodicy in Seventeenth-Century English Thought," *Journal of the History of Philosophy* 31.4 (1993): 519–44, traces the very different ideas about animal souls, from Descartes to panpsychism. Harrison's essay also traces the place of Descartes in English ideas, and this is the focus of Marjorie Nicolson's "The Early Stage of Cartesianism in England," *Studies in Philology* 26 (1929): 356–74, and G. A. J. Rogers's "Descartes and the Method of English Science," *Annals of Science* 29.3 (1972): 237–55. Wallace Shugg traces fictional representations of the beast-machine hypothesis in "The Cartesian Beast-Machine in English Literature (1663–1750)," *Journal of the History of Ideas* 29 (1968): 279–92. The literary is also the focus of Andreas Holger Maehle's "Literary Responses to Animal Experimentation in Seventeenth and Eighteenth-Century Britain," *Medical History* 34 (1990): 27–51. On visual art, see Nathaniel Wolloch, "Dead Animals and the Beast-Machine: Seventeenth-century Netherlandish painting of dead animals as anti-Cartesian statements," *Art History* 22.5 (1999): 705–27.

On the use of animals in science in the period, see Anita Guerrini, "The Ethics of Animal Experimentation in Seventeenth-Century England," *Journal of the History of Ideas* 50.3 (1989): 391–407; Anita Guerrini, *Experimenting with Humans and Animals: From Galen to Animal Rights* (Baltimore and London: Johns Hopkins University Press, 2003); Peter Harrison, "Reading Vital Signs: Animals and the Experimental Philosophy," in *Renaissance Beasts,* ed. Fudge, 186–207; Andreas Holger Maehle and Ulrich Tröhler, "Animal Experimentation from Antiquity to the end of the Eighteenth Century: Attitudes and Arguments," in *Vivisection in Historical Perspective,* ed. N. A. Rupke (London and New York: Routledge, 1990), 14–47; Wallace Shugg, "Humanitarian Attitudes in the Early Animal Experiments of the Royal Society," *Annals of Science* 24 (1968): 227–38; and Nathaniel Wolloch, "Christiaan

Huygens's Attitude Toward Animals," *Journal of the History of Ideas* 61.3 (2000): 415–32. Malcolm R. Oster's "The 'Beame of Diuinity': Animal Suffering in the Early Thought of Robert Boyle," *British Journal for the History of Science* 22.2 (1989), 151–80, offers an important overview of ideas about science and animals in the period and also a transcription of Boyle's letter containing his ideas and concerns about animal sentience and science. J. J. Macintosh offers an overview and interpretation of Boyle's ideas in "Animals, Morality and Robert Boyle," *Dialogue: Canadian Philosophical Review* 35.3 (1996): 435–72. On Marin Cureau de la Chambre, see Albert G. A. Balz, "Louis de la Chambre," *The Philosophical Review* 39.4 (1930): 375–97; and on William Cavendish, see Karen L. Raber, "'Reasonable Creatures': William Cavendish and the Art of Dressage," in *Renaissance Culture and the Everyday*, ed. Patricia Fumerton and Simon Hunt (Philadelphia: University of Pennsylvania Press, 1999), 42–66.

Conclusion

An important overview of liberal humanist ideas and their relation to earlier debates is offered in Tony Davies, *Humanism* (London and New York: Routledge, 1997). Stephen Greenblatt's *Renaissance Self-Fashioning From More to Shakespeare* (1980; repr., Chicago: University of Chicago Press, 1984) and Francis Barker's *The Tremulous Private Body: Essays on Subjection* (1995; repr., Ann Arbor: University of Michigan Press, 2002) both offer important—if anthropocentric—interpretations of Renaissance humanity.

On Sir Kenelm Digby, see Michael Foster, "Sir Kenelm Digby (1603–65) as Man of Religion and Thinker—I," *Downside Review* 106.362 (1988): 35–58, and "Sir Kenelm Digby (1603–65) as Man of Religion and Thinker—II," *Downside Review* 106.363 (1988): 101–25; also Jonathan Sawday, "The Mint at Segovia: Digby, Hobbes, Charleton, and the Body as a Machine in the Seventeenth Century," *Prose Studies* 6 (1983): 21–35.

For overviews of contemporary experiments on animal intelligence see, for example, the experiments recorded in Marion Stamp Dawkins, *Through our eyes only? The search for animal consciousness* (1993; repr., Oxford: Oxford University Press, 1998), and Lesley J. Rogers, *Minds of Their Own: Thinking and Awareness in Animals* (St Leonards, NSW: Allen & Unwin, 1997). For further discussion of historiographical issues concerning animals, see Erica Fudge, "A Left-Handed Blow: Writing the History of Animals," in *Representing Animals*, ed. Nigel Rothfels (Bloomington and Indianapolis: Indiana University Press, 2002), 3–18.

On actor-network theory, Bruno Latour's *We Have Never Been Modern* (1991), trans. Catherine Porter (Cambridge Mass.: Harvard University Press,

2001) and John Law's "Notes on the Theory of the Actor-Network: Ordering, Strategy, and Heterogeneity," *Systems Practice* 5 (1992): 379–93, are important statements. The theory and its relevance to the study of animals first came to my attention in Chris Philo and Chris Wilbert, "Animal spaces, beastly places: an introduction," in *Animal Spaces, Beastly Places: New geographies of human-animal relations,* ed. Philo and Wilbert (London and New York: Routledge, 2000), 1–34. Philip Armstrong, "What Animals Mean, in *Moby-Dick,* for example," *Textual Practice* 19.1 (2005): 93–111, also uses Latour's ideas in his analysis.

Index